增訂三版

國際商務契約

─實用中英對照範例集─

International Commercial Agreements

陳春山　著

三民書局

國家圖書館出版品預行編目資料

國際商務契約：實用中英對照範例集／陳春山編著.－
－增訂三版二刷.－－臺北市：三民，2016
面；　公分

ISBN 978-957-14-5741-3　(平裝)

1.商業應用文 2.契約 3.法律 4.國際商法

493.6　　　　　　　　　　　　　　101022800

© 　國際商務契約
　　　——實用中英對照範例集

編 著 者	陳春山
發 行 人	劉振強
著作財產權人	三民書局股份有限公司
發 行 所	三民書局股份有限公司
	地址　臺北市復興北路386號
	電話　(02)25006600
	郵撥帳號　0009998-5
門 市 部	(復北店)臺北市復興北路386號
	(重南店)臺北市重慶南路一段61號
出版日期	初版一刷　1996年8月
	增訂二版一刷　2001年6月
	增訂三版一刷　2013年1月
	增訂三版二刷　2016年10月
編 號	S 584520

行政院新聞局登記證局版臺業字第○二○○號

有著作權‧不准侵害

ISBN 978-957-14-5741-3　(平裝)

http://www.sanmin.com.tw　三民網路書店
※本書如有缺頁、破損或裝訂錯誤，請寄回本公司更換。

Giving the continuing threats
and opportunities nations
collectively face, it is clear that
people must think as globally
as possible.

Prof. Allen Sens, UBC

增訂三版序

　　臺灣發展唯一的路是──面向國際，臺灣社會經濟再升級的路──培育國際創新人才，過去六十年臺灣每一次有大幅社經提升，都是全球化升級，但總體而言，過去二十年，臺灣全球化進展不若英語系國家如香港、新加坡，離韓國也將顯現差距，這些都是臺灣不夠全球化所致。管理大師杜拉克曾述：全球經濟將取代個別經濟，臺灣必須在全球經濟展現利基、展現價值。實踐這個願景的三個策略是：全球性開放引進全球最好的產業（包括教育產業）、培育及引進全球化人才（技術及語言）、扶植全球品牌及創新產業，此三項策略以培育全球化人才為重中之重，全球化人才又以國際商務規劃及法務人才最重要，期以本書再版，對臺灣全球化人才培育，略盡棉薄之力。

陳春山

於國立臺北科技大學／智慧財產權研究所

2012 年 11 月

增訂二版序

　　隨著我國加入 WTO 的進展，我國企業以臺灣為營運中心 (Operating Center) 之需求亦漸為提高，我國企業不論於大陸、東南亞、中美洲設立製造或行銷據點，或與歐美國家進行合資 (Joint Venture) 或貿易，必須轉型為國際企業 (Multinational Enterprise)，而進行各類國際商業活動，此等活動需有完整之法務規劃 (Legal Planning)，針對不同之商業行為，諸如投資、買賣、僱傭等為法律分析，並研擬法律文件，本書此次再版即進一步擴充各種商務活動之契約範例，以期我國企業得以廣泛應用，並可為法律學院（系）教學之教材。

陳春山

於亞卓國際法律事務所／國立臺北大學法學院

2001 年 4 月

序

　　本書旨在供律師及會計師等專業人員，及商業人士了解國際商務契約之理論與實務，並得以實際運用於商業實務之範本。本書之內容包括：

　　1.契約書及契約法等基本概念，

　　2.契約書的形式及構造，

　　3.契約書的內容與基本條款，

　　4.買賣契約書的基本條款與範例，

　　5.其他各種契約書（經銷契約、租賃契約、貸款契約、聘僱契約、保證契約、股東合資契約、技術協助契約及不動產抵押契約等）的範例。前述範例並有中英對照以供參考。

　　以本書之篇幅，實無法包括商業活動中所需要之各種契約，但本書就契約書之架構有簡要且實際之說明，並提供各種範例，應可為其他各類型之契約之參考。

　　為便於商業人員與專業人士的使用，本書於每種契約範例之前，均列有各種契約之檢查表，讀者可於草擬契約或商業談判時，就該檢查表所列的事項一一加以檢視，如此之工作可避免當事人就重要事項未為訂定，而引起將來之爭議。

　　本書為供實務人士之用，故不強調法學理論之分析。又本書之說明仍以我國契約法為主，並輔以美國統一商法典 (Uniform Commercial Code) 之內容，理由乃因於我國進行之涉外交易，多以我國法為準據法，但仍不少以紐約法或加州法為準據法者，故有

參酌各州立法之基準法——美國統一商法典之必要。

不論讀者所從事的是製造業或服務業，亦不論所進行的交易是國內或涉外的型態，協商契約、草擬契約、訂定契約、解釋契約，甚至進行契約的訴訟，本書均為實用的參考資料，作者的另一本著作——契約法講義論，則進一步就各種型態的契約，有較深入的法律分析，供讀者參考。

本書乃作者利用課餘時間撰寫，因倉促付印，錯誤在所難免，尚祈讀者指正。又本書之所以得為出版，感謝中興大學法律系所諸教授、先進之指導，三民書局劉董事長及編輯同仁之支持，我的牽手月娥之協助繕打整理，藉此致謝。

陳春山

於中興大學法律系

1996 年 6 月

目次

第一篇

契約書及契約法的基本概念

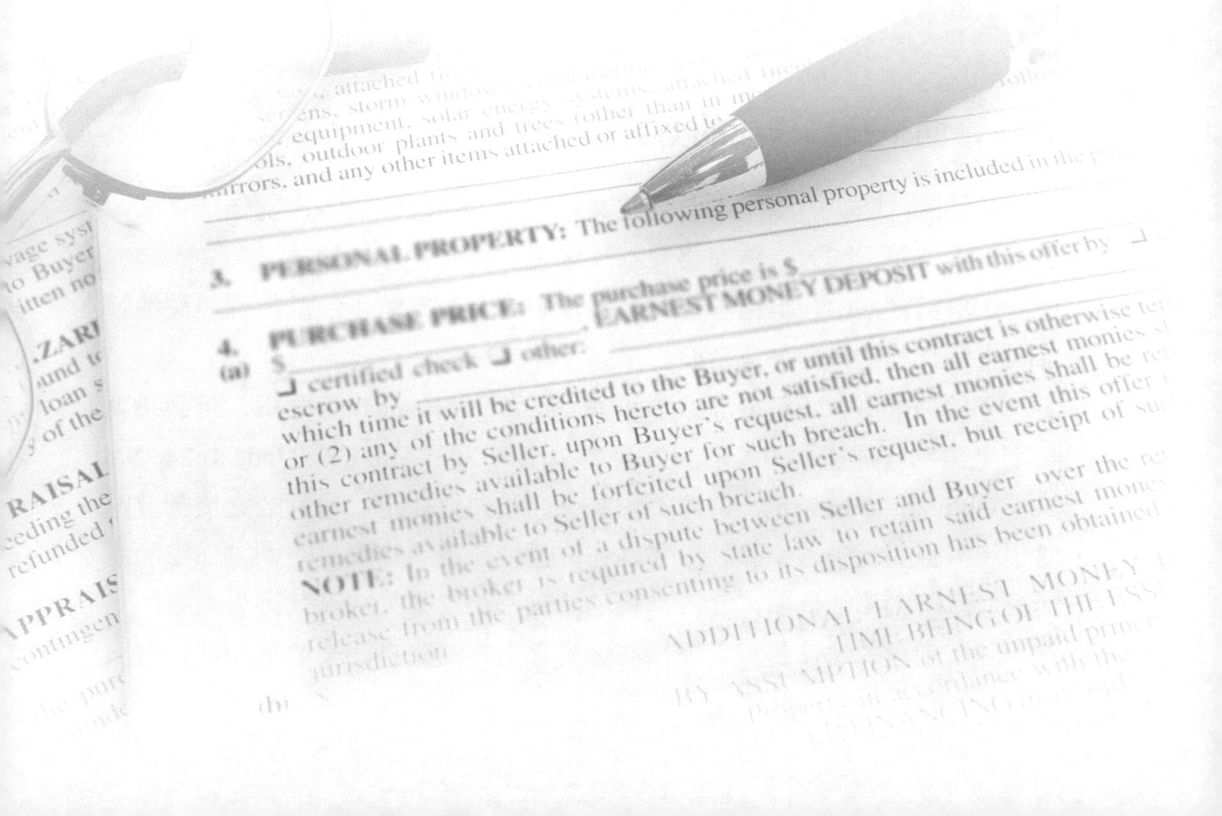

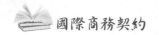

§1 國際商務契約書作成的必要性

1.「英文」商務契約的必要性

　　由於我國的官方語言為中文，因此，於臺灣地區所使用的契約文字，均以中文為主。然而，由於我國對外經貿活動之活躍，英文商務契約的利用，乃為商業人士所必需。尤其，我國正朝向經濟全面國際化，並已加入世界貿易組織，朝亞太營運中心的方向建立，亦為政府重大政策，因此，臺灣國際經貿活動的質與量將有迅速提升之趨勢。從事此等國際經貿活動，英文商務契約，乃不可避免。不論我國商業人士與英美法系國家，包括：美國、英國、新加坡、香港、加拿大、澳洲及紐西蘭等國家，當然須使用英文商務契約，即使與歐洲聯盟 (European Union) 等國家，甚至與其他亞洲國家，包括：日本及東南亞國家等，亦必須使用英文商務契約書，因此，英文商務契約乃我國商業及專業人士從事國際經貿活動所不可或缺之工具。

2.「書面」契約作成的必要性

　　於我國，除了特定所謂之要式契約之外契約關係並不需要作成書面，然於英美法系之國家，將商務契約作成書面乃特定法律的要求：

　　⑴於英國法中，英國於一九五四年制定有詐欺防止法，就保證契約及有關土地之契約，需作成書面。

　　⑵於美國法中，基於詐欺防止法 (Statute of Frauds) 的要求，乃必須就特定之契約作成書面 (UCC 1–206, 2–201, 9–203; Cal. Civ. Code 1624, NY Gen. Obligations Law 5)。根據美國統一商法典之規定，凡契約之價額或賠償超過五百美元者，須強制以書面作成且記載下列事項，否則，無法請求履行：其一、標的價額，其二、標的，其三、當事人或其代理人簽署。

　　除了前述的法律要求之外，商業實務上亦認定書面契約的作成，有利

於商業運作及當事人爭議的防免，以下將理由說明如下：

　　⑴當事人就其意圖有爭議或進行訴訟時，契約書得以為當事人合意的有利證據。

　　⑵英美法上有所謂之口頭證據原則 (Parol Evidence Rule)，該原則乃是主張當事人間就口頭或其他書面之溝通，以訂定正式的書面契約時，如該書面契約與其他先前之合意有不同時，應以該書面契約為準。因此，說書面契約有統合 (Integrate) 的效果。

　　⑶於英美法中，契約須有約因 (Consideration)，否則，該契約即為無效 (UCC 2–209)。書面契約就約因之認定有所助益 (UCC 2–205)。

　　書面契約之重要性就商業人士而言，乃在確定當事人之意思，而使商業運作更為順利，因此，我國專業及商業人士與外國廠商進行交易時，須盡量訂立書面契約，以保障自己之權益。

各國法規——Think as globally as possible.

Statute of Frauds (U.C.C.)

　　A contract for the sale of goods for the price of $5,000 or more is not enforceable by way of action or defense unless there is some record sufficient to indicate that a contract for sale has been made between the parties and signed by the party against which enforcement is sought or by the party's authorized agent or broker. (U.C.C. 2–201)

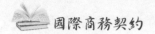

§2 契約書的形式

　　契約關係以書面作成者，較具證據力，就其書面亦得有兩種方式而作成，其得為正式契約書 (Formal Document) 及書信 (Letter Form) 的形式作成。一般而言，以契約書作成者，其為較重要或繼續性之契約；而以書信作成之契約其重要性較低。以正式契約書作成之契約文書，例如包括：外國之有價證券之承銷契約，而以書信方式為之者，例如：物品之買賣乃藉由買受及出賣之信函而成立買賣契約。然一般而言，仍以正式之契約書訂定契約較為常見且較適當。

§3　契約的生命週期

契約如同人的生命，皆有其期間、期限、生命週期，即契約的出生（即契約成立）、發展（即契約的效力與履行）、挫折（違約救濟）及死亡（終止、解除、消滅），民法總則、債編及其他特別契約法（如證券投資信託契約之特別法），皆對此有所規範，但除強制規定外，皆依契約自由原則運作。

1.是否成立契約?
  Is there a contract?
  ↓
2.契約是否有效?
  Is contract enforceable?
  ↓
3.誰可要求履行契約?
  Who can enforce contract?
  ↓
4.違約救濟?
  What are remedies?
  ↓
5.消滅
  Termination

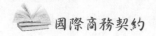

§4 契約的成立

契約為法律行為之一種，法律行為之成立要件，須具備當事人、標的及當事人之意思表示，而就契約而言，其成立必須意思表示合致，始可成立契約。惟應依何種方法當事人契約始可成立，而意思表示為合致，其應依下列之方法為之：

1.要約 (Offer) 與承諾 (Acceptance) 一致

即當事人於要約與承諾為明示或默示之意思表示，且其內容一致者，契約即為成立（民法第一五三條）。

2.交錯要約

即當事人之要約為偶然之一致，且其主觀之意思及客觀之內容一致，即成立契約。

3.意思實現

即依習慣或事件之性質，要約人於要約時聲明承諾無須通知者，或於相當期間內有可認為承諾之事實時，承諾即生效力，而使契約成立（民法第一六一條）。就懸賞契約是否為單獨行為或契約，學說上有爭執。但為符契約正義，並保障被懸賞人之權益，以單獨行為說較妥。

就英美法上買賣契約之成立，美國統一商法典第 2-204 至 2-207 條亦有特別之規定。

各國法規——Think as globally as possible.

Parol or Extrinsic Evidence (U.C.C.)

Terms with respect to which the confirmatory records of the parties agree or which are otherwise set forth in a record intended by the parties as a **final expression** of their agreement with respect to such terms as are included therein may not be contradicted by evidence of any prior agreement or of a contemporaneous oral agreement but may be supplemented by evidence of:

　　(a) course of performance, course of dealing, or usage of trade; and

　　(b) consistent additional terms unless the court finds the record to have been intended also as a complete and exclusive statement of the terms of the agreement. (U.C.C. 2–202)

各國法規──Think as globally as possible.

Creating a Contract (CISG)

■ A contract is concluded at the moment when an **acceptance** of an **offer** becomes effective in accordance with the provisions of this Convention. (CISG, 23)

■ Offer: (1) A proposal for concluding a contract addressed to one or more specific persons constitutes an offer if it is sufficiently definite and indicates the intention of the offeror to be bound in case of acceptance. A proposal is sufficiently definite if it indicates the goods and expressly or implicitly fixes or makes provision for determining the quantity and the price. (2) A proposal other than one addressed to one or more specific persons is to be considered merely as an invitation to make offers, unless the contrary is clearly indicated by the person making the proposal. (14)

■ Acceptance: (1) A statement made by or other conduct of the offeree indicating assent to an offer is an acceptance. Silence or inactivity does not in itself amount to acceptance. (2) An acceptance of an offer becomes effective at the moment the indication of assent reaches the offeror. (3)

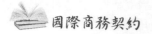

However, if, by virtue of the offer or as a result of practices which the parties have established between themselves or of usage, the offeree may indicate assent by performing an act, such as one relating to the dispatch of the goods or payment of the price, without notice to the offeror, the acceptance is effective at the moment the act is performed. (18)

各國法規——Think as globally as possible.

Formation of A Contract (U.C.C.)

■ Contract: A contract for sale of goods may be made in any manner sufficient to show agreement, including offer and acceptance, conduct by both parties which recognizes the existence of a contract, the interaction of electronic agents, and the interaction of an electronic agent and an individual.

■ Offer & Acceptance: (1) Unless otherwise unambiguously indicated by the language or circumstances: (a) an offer to make a contract shall be construed as inviting acceptance in any manner and by any medium reasonable in the circumstances: (b) an order or other offer to buy goods for prompt or current shipment shall be construed as inviting acceptance either by a prompt promise to ship or by the prompt or current shipment of conforming or nonconforming goods, but the shipment of nonconforming goods is not an acceptance if the seller seasonably notifies the buyer that the shipment is offered only as an accommodation to the buyer.

■ (2) If the beginning of a requested performance is a reasonable mode of acceptance, an offeror that is not notified of acceptance within a reasonable time may treat the offer as having lapsed before acceptance.

■ (3) A definite and seasonable expression of acceptance in a record operates as an acceptance even if it contains terms additional to or different from the offer. (U.C.C. 2–204)

各國法規──Think as globally as possible.

Implied Contract, Quasi Contract & Promissory Estoppel (Restatement)

■ Implied Contract: the surrounding facts and circumstances indicate that an agreement is reached

■ Quasi Contract: a legal fiction created by the court to avoid injustice/unjust enrichment

■ Promissory Estoppel: A promise which the promisor should reasonably expect to induce action or forbearance on the part of the promisee or a third person and which does induce such action or forbearance is binding if injustice can be avoided only by enforcement of the promise. The remedy granted for breach may be limited as justice requires. (Restatement S. 90)

§5 契約的效力與履行 (Effect and Performance of Contract)

　　於契約成立且生效後，當事人依契約之內容即需為給付，即債權人有請求給付之權利，而債務人有給付之義務（參考王伯琦：民法總則，第149頁）。所謂之給付義務，即為當事人間之權利義務關係，以買賣關係而論，出賣人需為財產移轉之給付，並負瑕疵擔保責任，而對買受人而言，其需為有關價金之支付及受領等給付義務。此種給付義務及內容，依契約自由原則，乃由雙方當事人加以規定，如當事人未為約定者，則依債法各論典型契約之規範而判定當事人之權利義務。再者，如依典型契約亦無從為判斷者，則依非典型契約之解釋方法，由法院依當事人之真意及契約真意為解釋之。

　　民法對於雙方當事人負有對價關係之雙務契約，特別規定其效力，其特殊之效力有二：即同時履行抗辯權及危險負擔。

1.同時履行抗辯權

　　所謂之同時履行抗辯權，乃是雙務契約之一方當事人，於他方未為給付前，得拒絕自己給付之權利（民法第二六四條）。行使同時履行抗辯權必須符合下列三要件：

　　⑴須雙方債務互為對價，

　　⑵須雙方各無先為給付之義務，

　　⑶須他方未依債之本旨為給付。

2.危險負擔 (Risk of Loss)

　　依契約成立生效後，發生契約之給付不能時，應依下列之規定辦理：

　　⑴如為可歸責於債務人之事由時，債權人得請求損害賠償（民法第二二六條），並得為解除契約及請求損害賠償（民法第二五六條、第二六〇條）。

(2)如為可歸責於債權人之事由時，債務人免給付義務，但仍得請求對待給付（民法第二六七條）。

(3)如不可歸責於雙方當事人者，免為對待給付之義務，如僅於一部給付不能者，按其比例減少對待給付（民法第二六六條）。

各國法規──Think as globally as possible.

Third-Party Obligatory/Beneficial Contracts (Taiwan)

■ One of the parties to a contract who has undertaken that an obligation shall be performed by a third party shall be responsible for the injury if the third party does not perform the obligation. (Taiwan Civil Code, 268)

■ When it is provided in a contract that an obligation shall be performed to a third party, the offeror may demand the debtor to perform the obligation to the third party, and such third party also has the right to demand performance direct from the debtor.

So long as the third party has not expressed his intent to take advantage of the contract specified in the preceding paragraph, the parties may modify the contract or revoke it.

If the third party expresses to either of the parties his intent not to take advantage of the contract, he is deemed to never have any right under the contract. (269)

各國法規──Think as globally as possible.

Effect and Performance of A Contract (Taiwan)

■ Performance

When performance has been made to the creditor or to his qualified representative in conformity with the tenor of the obligation, and has been accepted, the obligation is extinguished. (Taiwan Civil Code, 309 I)

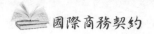

An obligation may be performed by a third party unless otherwise agreed by the parties or accorded with the nature of the obligation.

If the debtor objects to the obligation being performed by a third party, the creditor may refuse such performance; but if the third party has the interest of conflicts on the performance of the obligation, the creditor shall not refuse. (311)

■ Standards of Performance

The debtor shall be responsible for his acts, whether intentional or negligent. The extent of responsibility for one's negligence varies with the particular nature of the affair; but such responsibility shall be lessened, if the affair is not intended to procure interests to the debtor. (220)

§6　契約義務的違反與救濟 (Breach and Remedies)

契約已成立且生效後，當事人即負履行義務 (The Obligation of Performance)，當事人就債務不能履行或給付時，其構成債務之不履行，當事人即得依法請求救濟。按債務人不履行之狀況可分為下列幾種：不為給付、給付遲延、給付不能及不完全給付等情形。以下將就各種債務不履行之狀況為說明，並就其救濟之方式為概要之補充。

1.不為給付 (Repudiation)

所謂之不為給付即為給付之拒絕，乃給付可能而債務人表示不為給付之意思，按不為給付有認為即為給付遲延者，然給付拒絕可分為清償期前之給付拒絕及清償期後之給付拒絕,清償期後之給付拒絕應與給付遲延同，其救濟之方式亦與給付遲延同。而清償期前之給付拒絕，依民事訴訟法第二四六條之規定，其應構成有到期不履行之虞，即可於期前起訴。再者，債權人亦可拒絕受領而請求不履行之損害賠償 (Damages)（民法第二三二條、第二二七條）。又債權人亦可拒絕自己之給付，而行使所謂之不安抗辯權（民法第二六五條）。實務上常有給付拒絕之狀況發生，最高法院曾將不為給付解為給付遲延（七十八年臺上字第六三六號判決），而於履行期以前，因債務人責任尚未發生，無債務不履行之問題（八十三年臺上字第二四一〇號判決）。美國統一商法典 U.C.C. 第 2–610 條則訂有明確之救濟方式，聯合國貨物買賣公約第 71 條亦同。U.C.C. 第 2–610 條且對預期違約，債權人可請求之救濟為規範。

2.給付遲延

給付遲延可分成債務人及債權人之給付遲延，以下分別說明之。

債務人之給付遲延須符下列之要件：其一、須債務已屆給付期：如給

付定有期限者，債務人自期限屆滿時，即負遲延責任（民法第二二九條第一項）；如無定期限者，債務人於債權人得請求給付時，經其催告而未為給付，自受催告時起，負遲延責任（民法第二二九條第二項）。其二、其可歸責於債務人：可歸責於債務人，債務人始負遲延之責（民法第二三○條）。如債務人於履行期以前為拒絕給付，則因債務人責任仍為未發生，原則上無債務不履行之問題，如於履行期後為拒絕給付乃債務人給付遲延，與民法第二二七條無涉（八十三年臺上字第二四一○號判決）。於債務人遲延之狀況，縱發生遲延後之不可抗力因素所造成之損害，債務人亦須負其責任（民法第二三一條第二項）。債務人所應負責之形態如下：

(1)賠償因遲延所生損害

債務人遲延者，債權人得請求賠償因遲延所生之損害（民法第二三一條）。

(2)債權人得拒絕給付

對遲延之給付，如對債權人無利益時，債權人得拒絕給付，並得請求因不履行所生之損害賠償（民法第二三二條）。

(3)債權人得請求遲延利息

遲延之債務如以支付金錢為標的者，債權人得請求依法定利率計算利息，但約定利率較高者，仍從其約定利率。對於利息無須支付遲延利息（民法第二三三條）。

(4)債權人得催告其履行，如於期限內未履行者，得解除契約（民法第二五四條）

所謂之債權人遲延，乃是債務人已提出給付而債權人拒絕受領或不能受領之情形。構成債權人遲延必須符合下列要件：

①債務人已提出給付

即債務人已提出給付，而使債權人處於得以領受之地位，其提出得為現實之提出或言詞之提出。

②債權人拒絕受領或不能受領

即債權人明示拒絕受領，或債權人主觀上不能受領，如屬客觀不能受

領，應依不可歸責於因債務人之給付不能規定辦理（民法第二二五條）。

如符前述之要件者，即構成債權人之遲延，但給付無確定期限，或債務人於清償期前得為給付者，債權人就一時不能受領之情事，不負遲延之責任（民法第二三六條）。

就債權人之遲延，乃生下列之法律效果：

⑴債務人僅就故意或重大過失負其責任（民法第二三七條）。

⑵債務人無需支付利息（民法第二三八條）。

⑶債務人僅對已收取之孳息負返還之責任（民法第二三九條）。

⑷債務人得請求賠償因提出及保管給付物之必要費用（民法第二四○條）。

⑸債務人得依提存或拋棄占有之方式，而消滅其債務（鄭玉波：民法債編總論，第 176 頁；王伯琦：民法債編總論，第 307 頁以下）。

3. 給付不能 (Impossibility of Performing Condition)

給付不能可分為自始給付不能及嗣後之給付不能，自始之給付不能為契約是否具生效要件之問題，而自始之給付不能包括自始之主觀給付不能及自始之客觀給付不能。兩者之區別，依通說乃指其是否為任何人或僅對債務人為不能給付者。依學說之見解，自始主觀不能契約仍為有效，僅於自始客觀不能時契約為無效（民法第二四六條）。但此種客觀自始不能之情形可以除去者，且當事人締約時並預期於不能之情形除去後為給付者，其契約亦為有效（民法第二四六條第一項）。同理，附停止條件或始期之契約，於條件成就或期限屆至前，不能之情形已除去者，其契約亦為有效（民法第二四六條第二項）。對於自始客觀不能，得形成締約上之過失，非因過失而信賴契約為有效致受損害者，得向對造請求損害賠償（民法第二四七條第一項）。就嗣後之給付不能，其契約仍為有效，如不可歸責於債務人，債務人免負給付義務（民法第二二五條第一項）。如為可歸責於債務人，債權人得請求損害賠償（民法第二二六條），就雙務契約而言，其應依前述危險負擔之規定予以解釋之。

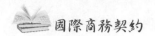

4.不完全給付

　　所謂之不完全給付乃是債務人未依債務之本旨而為給付者。不完全給付乃是債務人違反給付義務或附隨義務,而造成瑕疵之給付或加害之給付。就不完全給付之法律依據,民法債編總論已為修正,認其法律依據為民法第二二七條,不完全給付之法律效果,我國法有明文規定,僅得依給付遲延與給付不能之規定。

　　⑴債權人得為拒絕受領或要求補正（民法第二三五條）。

　　⑵債權人亦可請求損害賠償（六十六年臺上字第二一四二號判決）。

　　⑶又給付不完全得為補正者,因得適用民法第二五四條之規定,訂其期限補正,不補正者,得進而為解除契約。如不完全給付無法補正時,則得適用民法第二五六條之規定,予以解除契約。不完全給付與其他法律上之救濟相競合,例如不完全給付與瑕疵擔保責任常有競合之狀況,則債權人得分別依不完全給付或物之瑕疵擔保責任為分別之請求。如不完全給付對消費者構成安全或衛生上之危險,即為不完全給付與消保法請求權競合之適用問題（消費者保護法第七條以下）。

各國法規——Think as globally as possible.

Impossibility (Taiwan)

　　The debtor will be released from his obligation to perform if the performance becomes impossible by reason of a circumstance to which he is not imputed. If the debtor is entitled to claim compensation for the injury against a third party in consequence of the impossibility of the performance under the preceding paragraph, the creditor may claim against the debtor for the transfer of the claim for the injury, or for the delivery of the compensation he has received. (Taiwan Civil Code, 225)

　　If the performance becomes impossible by reason of a circumstance to which the debtor is imputed, the creditor may claim compensation for any injury arising therefrom. In the case specified in the preceding paragraph, if one part of the performance becomes impossible and the remaining part, if performed, will be of no interests to the creditor, the creditor may refuse the performance of the remaining part and claim compensation for the injury arising from complete non-performance. (226)

各國法規——Think as globally as possible.

Defaults (Taiwan)

■ When there is a definite period fixed for the performance of an obligation, the debtor is responsible for the default from the moment when such period expires. When there is no definite period fixed for the performance of the obligation, and when the creditor may demand the performance, but the debtor failed to perform the same after the creditor has notified him of the demand, the debtor is responsible for the default from the moment when he has been notified. The effect of instituting an action for performance and the service of the complaint, or the service of an order for payment according to the hortatory process, or any other similar act is equivalent to a notice. If there is a period fixed for the performance in the notice of the preceding paragraph, the debtor is responsible for the default from the moment when such time expires. (Taiwan Civil Code, 229)

■ When the debtor is in default, the creditor is entitled to claim compensation for any injury arising therefrom. So long as the default continues, the debtor under the preceding paragraph shall also be responsible for any injury arising from circumstances of force majeure, unless he can prove that the injury would have been sustained, even if he had performed in due time. (231)

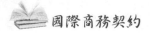 國際商務契約

各國法規——Think as globally as possible.

Incomplete Performance (Taiwan)

If a debtor incompletely performs his obligation by reason of a circumstance to which the debtor is imputed, the creditor may execute his right according to the provisions of the default or the impossibility of the performance.

In addition to the injury arising from the incomplete performance in the preceding paragraph, the creditor may claim compensation for other injuries arising therefrom, if any. (Taiwan Civil Code, 227)

各國法規——Think as globally as possible.

Excuses for Non-performance (Taiwan)

■ Illegality

■ Impossibility: If none of the parties is imputed to the impossibility of one party's performance, the other party shall be released from his obligation to perform the counter-prestation. If the impossibility is only partial, the counter-prestation shall be reduced proportionately. (Taiwan Civil Code, 266)

■ Default: The debtor is not being responsible for the default if the prestation has not been made by reason of circumstances to which he is not imputed. (230)

■ Unpredictable: If there is change of circumstances which is not predictable then after the constitution of the contract, and if the performance of the original obligation arising therefrom will become obviously unfair, the party may apply to the court for increasing or reducing his payment, or altering the original obligation.

The provision in the preceding paragraph shall apply mutatis mutandis to the obligation not arising from the contract. (227–2)

■ counter-prestation: A party to a mutual contract may refuse to perform his part until the counter-prestation has been performed by the other party, except he is bound to perform first.

When one party has partially performed his part, the other party shall not refuse his counter-prestation if circumstances are such that a refusal to perform would be against the manners of good faith. (264)

■ A person who is bound to perform his part first may, if after the constitution of the contract the property of the other party have obviously decreased whereby the counter-prestation might become difficult to be performed, refuse to perform his part, until the other party has performed his part or furnished security for such performance. (265)

■ Public Interests: A right can not be exercised for the main purpose of violating public interests or damaging the others.

A right shall be exercised and a duty shall be performed in accordance with the means of good faith. (148)

各國法規——Think as globally as possible.

Excuses for Non-performance (CISG)

A party is not liable for a failure to perform any of his obligations if he proves that the failure was due to an impediment beyond his control and that he could not reasonably be expected to have taken the impediment into account at the time of the conclusion of the contract or to have avoided or overcome it or its consequences. (CISG, 79)

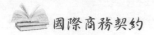

Anticipatory Repudiation & Insolvent (U.C.C.)

■ When either party repudiates the contract with respect to a performance not yet due the loss of which will substantially impair the value of the contract to the other, the aggrieved party may

(a) for a commercially reasonable time await performance by the repudiating party; or

(b) resort to any remedy for breach, even though he has notified the repudiating party that he would await the latter's performance and has urged retraction; and

(c) in either case suspend his own performance or proceed in accordance with the provisions of this Article on the seller's right to identify goods to the contract. (2–610)

■ Where the seller discovers the buyer to be insolvent he may refuse delivery except for cash including payment for all goods theretofore delivered under the contract, and stop delivery under this Article. (U.C.C. 2–702)

Anticipatory Repudiation (CISG)

■ A party may suspend the performance of his obligations if, after the conclusion of the contract, it becomes apparent that the other party will not perform a substantial part of his obligations as a result of:

(a) a serious deficiency in his ability to perform or in his creditworthiness; or

(b) his conduct in preparing to perform or in performing the contract. (71 (1))

■ If prior to the date for performance of the contract it is clear that one of the parties will commit a fundamental breach of contract, the other party may declare the contract avoided. (CISG, 72 (1))

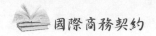

§7 契約關係的消滅 (Discharge of Contracts)

契約關係消滅之原因基本上可分為：契約之解除或終止、解除條件之成就、終期屆至、主體死亡及法律行為之撤銷等（鄭玉波：前揭書，第272頁以下）。我國民法對直接消滅之原因予以明示，即包括清償、提存、抵銷、免除及混同等原因。實務上常有爭議者，乃契約之解除及終止之事由。契約之解除與終止最大之不同，乃為契約之解除為自始之契約關係消滅，而契約之終止則為終止時起契約歸於消滅。契約之終止得依合意或法律規定而為終止，終止後之效果不得回復原狀，但仍得準用解除契約之相關規定（民法第二六三條）。

1.法定解除與終止 (Rescind & Terminate)

構成法定解除之原因包括：

⑴遲延給付

如遲延之給付對債權人仍有利益者，債權人得催告債務人履行，於期限內仍不履行時，債權人得解除契約（民法第二五四條），但如給付對債權人無利益時，債權人得逕行解除契約（民法第二五五條）。

⑵給付不能

因可歸責於債務人之事由致給付不能者，債權人得解除契約（民法第二五六條）。

⑶不完全給付

如不完全給付而不能補正時，則應類推適用民法第二五六條之規定得以解除契約。

2.約定解除與終止

所謂之約定解除，乃是依契約之規定，雙方保有解除權，依據此等契

約規定而得行使解除權，因而解除契約而消滅法律關係（我妻榮：債權各論，第 209 頁以下）。約定解除乃由當事人對他方為意思表示，其法律效果與法定解除相同。

3.合意解除

即於無法定事由及契約規定之狀況下，當事人是否得以任意約定解除契約，法律上尚有爭議。然依契約之自由原則，除法律另有規定者外，應准予當事人合意解除契約（我妻榮：前揭書，第 213 頁以下）。

契約解除後，當事人需依民法第二五九條規定回復原狀，且當事人得依據法定事由或約定事由請求損害賠償（民法第二六〇條）。再者，解除契約後之回復原狀，應適用雙務契約之同時履行抗辯權及危險負擔之規定（民法第二六一條）。

各國法規——Think as globally as possible.

Rescind & Terminate (Taiwan)

■ Rescind for Default: When a party to a contract is in default, the other party may fix a reasonable period and notify him to perform within that period. If the party in default does not perform within that period, the other party may rescind the contract. (Taiwan Civil Code, 254)

■ Rescind for Impossibility: In cases provided by Article 226, the creditor may rescind the contract. (255)

■ Deadline for Rescission: If there is no deadline for the exercise of the right of rescission, the other party may fix a reasonable one and notify the party having the right of rescission to make a definite reply within such deadline whether he will rescind the contract or not. If the notice of rescission is not received before such deadline, the right of rescission is extinguished. (257)

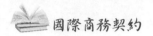

■ Notice for Rescission: The right of rescission shall be exercised with the expression of intent to the other party. (258)

■ Termination: The provisions of Articles 258 and 260 shall be mutatis mutandis applied when the parties terminate the contract in accordance with the provisions of the act. (261)

各國法規──Think as globally as possible.

Terminate (CISG)

■(1) A contract may be modified or terminated by the mere agreement of the parties.

■(2) A contract in writing which contains a provision requiring any modification or termination by agreement to be in writing may not be otherwise modified or terminated by agreement. However, a party may be precluded by his conduct from asserting such a provision to the extent that the other party has relied on that conduct. (CISG, 29)

各國法規──Think as globally as possible.

Effects of Rescission & Termination (Taiwan)

■ Legal Effects: Unless otherwise provided by the act or by the contract, each party shall, in case of rescission, restore the other party to his status quo ante according to the following rules:

(1) Each party shall return the prestation received to the other party.

(2) If the prestation received consisted of money, interest calculated from the time of receipt shall be added.

(3) If the prestation received consisted of service or of the use of a thing, the value of such service or use at the time of receipt shall be reimbursed in money.

(4) If a thing to be returned has produced profits, such profits shall be returned.

(5) If necessary or beneficial expenses of the thing to be returned have been paid, such expenses may be claimed for to the extent to which the other party is benefited at the time of return.

(6) If a thing to be returned has been damaged or destroyed or cannot be returned owing to any other cause, its value shall be reimbursed. (Taiwan Civil Code, 259)

■ Compensation: The exercise of the right of rescission does not prejudice to the claim for compensation. (260)

第二篇

契約書的形式及構造

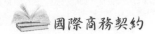

§1 契約書的整體形式

契約書於標題以下，分別按順序為前言、契約本題、末尾條款、署名及蓋印等項，以下乃就其形式上之結構圖示說明如下：

SALES AGREEMENT	標題
THIS AGREEMENT, entered into in (place) on the day of ＿＿＿＿, 2000, by and between A and B....	當事人及住所
WITNESSETH:	
	前言
WHEREAS ＿＿＿＿ ; and WHEREAS ＿＿＿＿ . NOW, THEREFORE, in consideration of the premises ＿＿＿＿＿＿＿＿＿＿＿＿＿＿＿＿＿＿＿＿＿＿＿ .	說明條款 約因
IT IS AGREED:	
1. ＿＿＿＿＿＿＿＿＿＿＿＿＿＿＿＿＿＿＿＿＿ . 2. ＿＿＿＿＿＿＿＿＿＿＿＿＿＿＿＿＿＿＿＿＿ . 3. ＿＿＿＿＿＿＿＿＿＿＿＿＿＿＿＿＿＿＿＿＿ .	合意之內容
IN WITNESS WHEREOF, the parties hereto have on the day and year first above written caused these presents to be executed in their behalf and in their corporate names—respectively by their proper officers hereunto duly authorized and their respective corporate seals to be hereto attached by	末尾條款

like authority.

Corporate Seal	_____ CORPORATION	署名蓋印
ATTEST	By _____	
	President	
_____		
Secretary		
Corporate Seal	THE _____ COMPANY	
ATTEST	By _____	
	Managing Director	
_____		
Secretary		

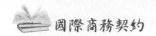

§2 訂約的開始及年月日

　　契約之開頭須列明該文件為契約，且須記載當事人之姓名及地址，就當事人之名稱得記載其於契約中之地位，例如：賣方 (Seller) 或買方 (Buyer) 等，如無適當之名稱描述當事人者，則須將當事人於後述之條款中，予以重述。

　　於契約中所使用之語言及文字須避免引起爭議，尤其當事人之一方為消費者時，美國之州法中乃要求其契約須為淺顯之文字 (Plain Language)。

　　就契約年月日之記載，通常均為如下之記載：

　　於＿＿＿年＿＿＿月＿＿＿日，A 及 B 之間訂立契約。

　　AGREEMENT made this ＿＿＿ day of ＿＿＿ between A and B.

　　如當事人欲使該契約溯及生效者，則非使用今日 (This) 而使用當時 (As of) 如此記載時，契約之效力乃以該記載日開始生效。其記載之方式如下：於年月日當時，A 與 B 訂定契約。

　　於＿＿＿年＿＿＿月＿＿＿日當時，A 及 B 間訂立契約。

　　AGREEMENT made as of ＿＿＿ day of ＿＿＿ between A and B.

---

§3　訂約的場所

　　由於國際商務英文契約之中，如當事人未訂有準據法，其訂定之場所，有助於準據法適用之認定，因此，實務上常於契約書最初部分及於訂約日之後，記載締約之場所，如未於該處記載者，則於契約成立證明之文句 (In Witness Clause) 之中為記載。

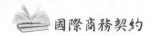

§4　當事人

　　為表明當事人之身分，如當事人為自然人者，須記載當事人之姓名及其住所。如當事人為法人或公司者，須表明公司名稱 (Corporate Name)，如公司有數營業處所者，應記載其主要之事務所，其通常之記載如下：A 公司為紐約州之公司，其主要之營業處所於紐約州紐約市　　街　　號。

　　為便於契約條款中提及當事人之名稱，乃於契約之最初文句中記明當事人名稱之縮寫，其記載常為 Hereinafter referred to as（以下簡稱）。

　　為避免引起爭議，契約條款中應盡量避免使用第一當事人或第二當事人之記載 (Parties of the First Part)，應記載其身分較為妥當，例如：買方 (Buyer) 或賣方 (Seller)。如不利於使用縮寫時，亦得使用略稱，例如 IBM 或 AT&T 等。

　　依前所述，契約之最初記載，得用正式條款或書信，以下乃將其形式提供範例說明如下：

　　THIS AGREEMENT, made this 　　 day of 　　 , 2000, by and between A COMPANY, INC. hereinafter called "A 　　 ", a corporation organized and existing under the laws of the state of 　　 , U.S.A. with its principal place of business at 　　 , U.S.A. and Y COMPANY, LTD. hereinafter called "Y 　　 ", a corporation organized and existing under the laws of the Republic of China with a principal place of business at 　　 , Taiwan.

---

§5　說明條款 (Whereas Clause)

說明條款通常乃以英文 Whereas 開始,說明條款為當事人目的之記載,此為英美法之法理，然於中文契約書中，並無記載此部分。就此說明條款之法律效力，除例外之情形外，該說明條款並無特別之法律效力，如欠缺記載該說明條款者，契約仍為有效。相反地，如該說明條款有特定法律效力時，就其記載之內容，應審慎為之。

說明條款就下列情形，有一定之法律效力：

⑴如契約之本體及內容記載欠缺明確，及無法由契約內容了解當事人之權益關係者，則得由說明條款之內容，用於推演當事人之爭議。

⑵依據禁反言之原則 (Estoppel by Representation)，如說明條款有記載之事項，當事人不得於訴訟上主張相反於該事實之內容。由此觀之，說明條款雖非契約之內容，然於解釋契約內容時，其有重要之意義。

就說明條款之內容，以下乃列舉範例供為參考：

WHEREAS, X ＿＿＿ has been, and is engaged in the business of manufacturing, buying, selling and generally dealing in ＿＿＿ ;
and

WHEREAS, X ＿＿＿ is willing to enter into a technological assistance agreement with such joint enterprise;

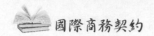

§6 約因 (Consideration)

於說明條款之後，常記載當事人所合意之交易約因，於英美法中，如無交易之約因則無法成立有效之契約，所謂之約因乃是指當事人間之法律拘束或法律上之利益。

於某些狀況下，各州並不要求須有約因之記載，或者對約因並無明示之表示，而須待證明者，然為避免爭議，且於通常之狀況下，契約中應明示當事人交易之約因，以下乃就約因之記載範例說明如下：

Now, therefore, the parties, each in consideration of the covenants and agreements of the other contained in this agreements, have consented and agreed as follows:

---

§7　合意的內容

　　契約書之說明條款及約因之後，即為契約之主要部分，該主要部分應記載當事人合意之內容，如其契約之內容過長者，則得分章節條款予以區分。就契約內容所記載之文字而言，為避免爭議，當事人應使用簡單及明瞭之用語，如該用語有諸多解釋之可能者，應對該用語予以定義。

　　就契約條款之內容，如當事人文句有所不明確者，則須契約條款之解釋，於英美法上，有所謂之口頭證據原則 (Parol Evidence Rule)，即依口頭證據之內容予以補充契約之內容，而於我國契約法之解釋，亦須探討客觀之證據，以探求當事人之真意。

　　由於契約書之內容，乃依各種契約條款有所不同，為通常契約之條款應包括下列四大類之內容：

1.基於契約應予履行之債務內容

　　⑴各種契約之共同條款
　　①定義條款
　　②契約期間之條款
　　③通知條款
　　④最終條款
　　⑤債權讓與及債務承受條款
　　⑵各種契約之特殊條款

2.契約履行所生問題之處理條款

　　⑴契約解除之條款
　　⑵契約修正或變更之條款
　　⑶損害賠償或違約金條款

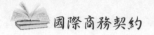

3.因情事變更之處理條款（不可抗力條款等）

4.糾紛解決條款

　　⑴裁判管轄條款

　　⑵準據法條款

　　⑶仲裁條款

　　就前述契約書之內容乃於第三篇有關契約書內容之基本條款予以詳細說明之。

§8　末尾條款 (Testimonium Clause)

於契約之最後有所謂之末尾條款或誓約條款，其乃記載當事人就契約之訂定得為證明，以下乃就該末尾條款之範例說明如下：

In Witness Whereof, the parties hereto have executed this agreement the day and year first above written.

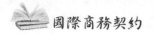

§9 署 名

　　於英美法實務中，當事人之簽名或其他記號，均足以證明當事人就契約為同意，然通常則以簽名之方式為之。然如公司之章程或法規要求者，則須加蓋公司印章。就以公司為署名者，該署名須有適當之授權，因此，於公司之名稱下方，須就該代表人之職稱予以記明，再者，為證明該公司印章使用之適法性，除記載該公司之代表人及姓名外，亦須就該法人之秘書 (Secretary) 之姓名，予以載明，該公司秘書乃保管公司印章及議事錄之記載人員，故得就該公司印章或代表人之簽名予以證明。

§10　公證人認證 (Acknowledgments)

　　一般契約並無須公證人 (Notary Public) 之認證，但如法規要求者不在此限。再者，就重要之契約書，如經適當之公證人認證，如契約當事人提起訴訟時，則對契約書內容得為有效證明。就契約書認證之文句範例說明如下：

CERTIFICATE OF SIGNATURE

We hereby certify that the signature of the undermentioned person, affixed to the accompanying document, is genuine and authentic.

Name of Person: _____
Title: _____
Name of the Company: _____

The R.O.C. CHAMBER OF COMMERCE

_____
Managing Director

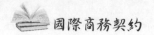

§11 刪除與修正

　　如契約經記載後須予以刪除或修正者，須為一定之記載，如為刪除者，需於刪除部分畫兩條平行線，而為明確之刪除，如僅為變更者，則須於刪除之部分上端為新內容之記載，惟不論刪除或變更，必須由當事人就其姓名 (Initial) 為署名，否則，不生其效力。

第三篇

契約書內容的基本條款

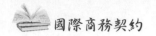

§1　定義條款

就契約有關之爭議，文義之不確定常為爭議之導火線，為避免此等爭議，於契約中須就特定之文句予以定義，該定義條款可達下列之功能：

⑴為避免契約文句過於冗長，如就特定之文句予以定義，得避免重複使用該文句。

⑵為避免同一文句重複使用時，其使用之內容與目的有所差異，乃將該文句予以一貫之定義，則得使該文句之使用具有一貫性。就定義條款之適用，舉例說明如下：

於本契約中，下列之名詞有以下之意義，但如該契約之內容有其他之意義者不在此限。

In this Agreement, the following terms having meaning unless the context clearly requires otherwise:

---

§2　契約存續期間

　　契約乃於契約期間而有其效力，該契約之時期或末期，乃依契約之記載而定，但如契約未記載者，得依契約之解釋而推測當事人之意思而定之，如契約記載訂定日者，當於該訂約之日起為效力發生之日。同理，契約於約期屆至時而消滅，如未訂有約期者，應依契約之解釋而推定當事人之意思。

　　就契約存續期間之記載，得依下列模式而記載其期間：

　　本契約經中華民國政府核准始，該契約期間為五年。

　　A 於期間屆滿前十二月，得通知 B 延長該契約之期間五年。

　　This agreement comes into force on the date which it is formally approved by the R.O.C. Government and continues for five years.

　　On the expiring of the period, A may continue this agreement for further five years by twelve months' prior written notice to B.

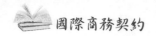

§3　通知條款

於契約生效後，契約當事人對他造為各項通知，就通知之問題包括通知之時期、對象及方法等須於契約中予以明訂。以下分別說明之：

1.通知之時期

契約條款往往就個別之事項有通知之時期，例如有關契約期限之延長，得約定於契約終了前十二個月內為之；或對有關董事會之召開，須於召開前將召集事由於開會前二十天內予以通知等。

2.通知之對象及地址

就通知之對象如當事人未有約定者，即以簽約人或簽約代表人為通知對象，如通知之對象為大型企業者，須就通知之人員於契約中予以訂定。再者，當事人亦得約定以他造之顧問律師為通知之對象。就契約當事人之地址須於契約中予以明訂，如當事人之地址有所變更時，於當事人通知對造後，該對造始有以新地址為通知之義務。

3.就通知之方法當事人得為約定

例如：得以傳真、快遞、空運或海運之方式而為送達，就上述之方式如為一般空運者，得要求掛號。

4.通知之生效

通知之生效時期得由當事人予以約定，如未為約定者，應其準據法而訂其生效時期，就英美法而言，其乃採發信主義，即於發信時，該通知即為生效。而我國民法則採到達主義，如擬改採發信主義者（即發信即生送達效力），須於契約中明訂之。

就一般契約有關通知條款所定之文句列舉範例如下：

本契約有關之通知、要求、同意、要約或請求等事項，必須以書面且以親自送達或以掛號、或傳真並掛號之方式向以下之地址為送達：

致：臺灣塑膠股份有限公司

　　（地址）

通知於發出後生效，就變更地址時，於該變更地址之通知到達對造後，該變更始生效力。

Any notice, reguest, consent, offer or demand required or permitted to be given in this agreement, must be in writing and must be in sufficiently given if delivered in person or sent by registered airmail or by cable confirmed by registered airmail letter, addressed as follows:

To: Taiwan Plastic Co., Ltd.

　　(address)

Notice must be deemed to have been given on the date of mailing except the notice of change of address which must be deemed to have been given when received.

§4　讓與或承擔條款

　　契約義務之讓與包括債權之讓與與債務之承擔，就我國法而言，除因債權之性質或特約債權禁止扣押等情形外，債權人得將債權讓與於第三人（民法第二九四條），就債務承擔而言，則須經債權人承認，始得由第三人承擔債務（同法第三〇一條）。由以上之規定可得知，債權原則上得為讓與，但債務之承擔非經債權人之同意，不得由第三人承擔之。惟當事人有特別約定者，如當事人認為契約關係乃因信賴關係而形成，故契約非經當事人同意而為讓與，基於此當事人之信賴關係，契約當事人得就讓與條款為以下之約定：

　　非經契約當事人對造之書面同意，任一方不得將契約之權利或責任讓與第三人。

　　Neither party shall, without the prior written consent of the other party, assign any of its rights and duties under this contract to any third party.

　　除一般契約權利與義務之轉讓外，合資契約有關股權或其他權利之轉讓，當事人以經由特別之約定，以限制其轉讓，以下乃以範例說明如下：

　　本契約非經他方之事前書面同意，不得轉讓予他人。

　　本公司之股份或權益，非經他造之優先認購，不得轉讓、設質、擔保、出賣或移轉於他人。該優先認購之要約須以書面且以掛號空運文件送達於其他股東。

　　於該出售之要約送達後六十天內，受要約人須以書面之航空郵寄予

要約人，對其要約是否接受或拒絕。如要約被拒絕或受要約人並無書面表示接受或拒絕，要約人得於接受拒絕之通知之日起三十日內，或於前述發通知之日起六十日內，對其股份為轉讓。如其要約經受要約人接受時，該股份應售予承諾人，且除非當事人另有合意，該股份應以書面承諾時之公司發行股份所得價值之兩倍，為要約人之售價。該公司之淨值應由 A 公司依一般會計原則決定之，A 公司之決定為終局且拘束當事人，該轉讓所生之費用應由要約人及承諾人平均分擔之。

This Agreement shall not be subject to assignment by either party without the prior written consent of the other.

None of the shares of stock of the COMPANY nor any interest therein shall be tran sferred, pledged, hypothecated, sold, assigned or in any way encumbered by either party, unless such offer shall be in writing and shall be delivered to the other shareholders by registered airmail.

Within sixty (60) days after delivery of such offer of sale, the offeree shall accept or reject such offer, in writing and by registered air mail. In the event of such offer being rejected, or in the event of no written acceptance or rejection being sent by the offeree, the offeror shall be free to transfer or otherwise dispose of such shares to others within thirty (30) days after the notice of rejection is received or after the expiration of the above sixty (60) days period. In the event of such offer being accepted, the shares so offerred shall be sold to the offeree and the price of each shares so sold shall be twice the amount obtained by dividing the net assets value of the COMPANY at the time the written acceptance is received by the offeror by the number of the then outstanding shares, unless otherwise agreed upon between the parties. The net assets value shall be determined by A, in accordance with the generally accepted accounting principles and the determination of A shall be final and binding on the parties. The fees of A shall be shared equally between the offeror and offeree.

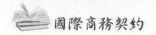

§5　履行條款

　　契約成立後，當事人即須依據契約履行 (Performance)，以買賣契約而言，買賣雙方乃就物品之出賣與買受為合意，其同意之條款如下：依據以下之條款賣方同意售予買方而買方同意至賣方購買物品。

　　就勞務提供契約而言，勞務提供者同意履行勞務，而對造則同意受領勞務並支付對價。

　　於契約發生債務不履行時，應依其不履行之狀況而由當事人負其責任。就履行之遲延而言，其得請求之權利，包括拒絕受領、請求損害賠償或解除契約等。就履行遲延之法律效果而言，為求明確，通常將履行之時期之重要性及效果予以特別之規定，試列舉範例如下：

　　如賣方未能依本契約之條款交付物品時，應按時交付為契約之重要本質，買方得為有權拒絕受領物品，且解除本契約之全部或一部。

Time of delivery is of the essence of this Contract, Buyer reserves the right to refuse any goods and to cancel all or any part of this Contract if seller fails to deliver any part of the goods in accordance with the terms of this Contract.

　　就損害賠償而言，為避免損害賠償額確定之困難，當事人得約定違約金 (Penalty)，該種違約金即為意定之損害賠償金 (Liquidated Damages)。

§6　責任限制及不可抗力條款

債務人因善意且無過失所生之債務不履行，尤其因不可抗力之事項，當事人乃合意免除其責任，有關不可歸責當事人所生之債務不履行，我國法乃規定債務人免給付義務（民法第二二五條第一項）。惟為避免爭議，尤其有關國際交易因各國法律之不同，未必有當然之法律效果，因此，乃就不可抗力之事由，規定當事人免給付之義務。就該種不可抗力條款之規定，試舉範例如下：

不可抗力

本契約當事人之任一方，就因下列事由所生之債務不履行不負其責任：火災、水災、罷工、勞動爭議及其他僱傭之紛爭、不可避免之事故、經宣告或非宣告之戰爭、禁運、封鎖、法律限制、騷擾、內亂、及其他非當事人所得控制之事由等。

FORCE　MAJEURE

Neither shall be liable for failure to perform, its part of this Agreement when such failure is due to fire, flood, strikes, labor troubles or other industrial disturbances, inevitable accidents, war (declared or undeclared), embargos, blockades, legal restrictions, riots, insurrections, or any cause beyond the control of the parties.

該種條款即所謂之自然現象 (Act of God) 或不可抗力 (Force Majeure)，該條款含有一般包括之文字，即屬同一種類 (Genus) 之事由者，亦得為不可抗力之事由，而使當事人免負其責任，該原則即所謂同種之原則 (Rule of Ejusdem Generis)。

§7 競業禁止條款

競業禁止條款包括營業受讓及員工之競業禁止條款，營業讓與之禁止條款乃指於營業讓與之後，為避免讓與人與受讓人形成競業之關係，因此，乃規定讓與人及讓與人之主要股東，不得從事同種之營業，於必要時，亦得就其競業禁止之義務，由受讓人向讓與人交付一定之對價(Consideration)。

就受僱契約或僱傭契約而言，為避免受僱人利用營業秘密，乃規定僱傭契約終了後之一定期間內，受僱人不得任職於同種類之企業。以下乃列舉範例如下：

受僱人乃同意於僱傭關係終止後兩年內，不論其終止是否具合法之理由，該受僱人不得就職於經營同種業務之公司，該受僱人亦不得以直接或間接之方式與本人、代理人或受僱人之身分，與公司進行同種之業務。當事人乃同意受僱人於違反本契約時，該公司得採取必要之法律救濟方法，向有管轄權之法院請求受僱人禁止該違反本契約之行為。

The employee agrees that for a period of two years following the termination of his employment, whether such termination be with or without cause, he will not enter the employment of any person, firm or corporation engaged in a similar line of business in competition with the Corporation, nor himself engage during such period, directly or indirectly, as principal, agent or employee, in any such business in competition with the Corporation. It is agreed that any breach of this agreement by the employee shall entitle the Corporation in addition to any court of competent jurisdiction to enjoin any violation of this agreement.

---

§8　仲裁條款

　　英文契約書中須就將來當事人之爭議之防免，而加以規定。就紛爭之解決方法得以仲裁或訴訟之方式為解決。仲裁比諸訴訟有下列之優點：

(1)由專門之仲裁人之仲裁判斷，其結論較為實際。

(2)仲裁之判斷其執行較為容易。

(3)一般而言，仲裁判斷之結果預見可能性較高。

(4)仲裁無須遵守審判公開之原則，得保持當事人之秘密。

(5)一般而言，仲裁之費用較為節省。

(6)仲裁之程序較一般訴訟為迅速。

　　國際商務仲裁須經當事人之同意始得為利用，故就仲裁之方法須為書面之合意（參考我國仲裁法第一條）。

　　就國際商務仲裁之條款，得依國際商務仲裁協會或其他外國或本國之仲裁規則予以決定，以下乃略舉美國商務仲裁協會所常用之標準條款說明如下：

　　因本契約所生之爭議，應依美國商務仲裁協會之規則，由單一或數仲裁人予以終局之決定。

　　All disputes arising in connection with the present contract shall be finally settled by arbitration in accordance with the Commercial Arbitration Rule of the American Arbitration Association by one or more arbitrators.

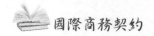

§9 程序法及管轄法院之條款

　　當事人之交易爭執除以仲裁解決外，亦得以訴訟方式為解決，以訴訟方法為解決者，即須先決定管轄法院，於決定管轄法院時，須考量管轄法院所為之裁判，是否得於當事人有財產之處獲得執行，再者，執行管轄法院時，亦須對準據法為考量，於通常之情形，管轄法院與準據法為同一，以避免管轄法院就外國法律之適用，而無法獲得相當之預測可能性。以下乃就管轄法院與準據法之標準條款說明如下：

　　本契約乃於英國作成並以英國法為準據法，並依英國法而為解釋且有其效力，當事人且同意以倫敦高等法院為管轄法院。

　　This Agreement must be construed and take effect as a Contract made in England and in accordance with the laws of England and the parties hereby submit to the jurisdiction of the English High Court of London.

　　當事人亦得就訴訟或裁判之管轄與準據法分別訂定條款，專就準據法訂定條款者，乃列舉範例如下：

　　本契約之效力、內容及履行，乃以紐約法為準據法，並依同法而為解釋之。

　　The validity, construction and performance of this agreement shall be governed by and interpreted in accordance with the laws of the State of New York.

---

§10　契約關係的消滅

於我國法中，契約關係之消滅包括契約之解除與終止，解除契約與終止契約最大之不同乃其法律效果，即當事人是否有回復原狀之義務。然該種區別，於英美法中較不明確，於英美法中得解除契約之情形，包括履行之拒絕 (Repudiation) 及重大契約之違反 (Fundamental Breach) 等，於契約解除後，當事人即無履行之義務，然就履行之部分，於英國法中，該消滅乃對將來之債務，並無回復原狀之問題，而於美國法中，則承認當事人可請求當事人回復原狀，於選擇準據法須就該區別有所認識，如依英國法時，當事人須就契約解除後之回復原狀予以特別規定，始有該種法律效果。契約之解除得因當事人之債務不履行而發生，其適用之標準條款如下：

如當事人之任一方有違本契約之條款，或其成為支付不能、破產、為債權人之利益轉讓本契約或設定信託擔保或進入清算程序，對造得以十日之通知解除本契約。

Upon a breach of any of the terms or conditions of this agreement by either party or should either party become insolvent, bankrupt, make an assignment or trust mortgage for the benefit of creditors, or enter into a receivership, this agreement may be terminated at the option of either party by writing ten days notice to the other party prior (date).

第四篇

買賣契約

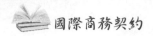

§1 序 說

　　買賣為今日私法中交易法之中心，而私法乃採取私法自治原則，為現代法之基本原理，私法自治之體現，即為契約自由原則，惟有契約自由原則，方可達成前述之經濟及文化效用。由此觀之，買賣契約關係之成立與內容，多隨當事人加以設定，其與物權契約顯有不同。然而，為達成公益之目的，對契約之自由多加限制，於戰時，國家往往對買賣之價金與貨物之供給加以限制，例如：國家總動員法為集中運用全國之人力、物力，對物質之使用及交易之價量等，均加以管制（參考我妻榮：前揭書，第 241 頁以下）。於平時為實現契約正義並保障經濟之弱者，對買賣契約之內容及成立亦多所限制，例如：消費者保護法對定型化契約、郵購買賣及訪問買賣等買賣契約，均有所限制（消費者保護法第一一條以下）。再者，買賣法對當事人之約定，應屬補充及任意性質，其原則上縱與買賣法之規定相衝突，亦非無效。但為保障公序良俗，例如：當事人得以特約免除或限制出賣人關於權利或物之瑕疵擔保義務（民法第三六六條），但如出賣人故意不告知瑕疵者，其特約即為無效。由此觀之，買賣法之規定，並非當然屬任意法。

§2　買賣契約的成立

　　契約之成立方式，應以當事人互相表示意思一致，契約方始成立（民法第一五三條第一項），就買賣契約而言，依民法第三四五條之規定，當事人必須就標的物及其價金，互相同意，買賣契約始為成立（民法第三四五條第二項），學者由比較民法第一五三條及第三四五條第二項之規定，認為現行法價金與標的物意思合致，即認為契約成立，有違契約自由原則，例如買賣雙方先就價金與標的物意思表示合致，然就清償地與清償時之問題，雙方認為應就此等事項另行合意，契約始為成立，以此種狀況下，不宜認為當事人就必要之點已經合致，而認為契約已經成立，應適用民法第一五三條之規定，認為僅對必要之點意思一致，而推定買賣契約成立，而不得立即視為契約已經成立（黃茂榮：買賣法，第 124 頁）。

　　就價金之部分，當事人得約定依市價訂之，該市價應解釋為標的物清償時清償地之市價，惟當事人亦得另為約定（民法第三四六條第二項）。如當事人並無特約，且並無約定市價者，買賣契約即可能不為成立，然實務認為雙方對公產為交易，其雖未約定價額，然如市場上有其一定價額，乃屬依其情形可定其價金，故亦認為契約成立（五十五年臺上字第一六四五號），此解釋應可認為乃民法第三四六條第二項之例外解釋，不宜認為原則之解釋。

　　各國對契約之成立要件均有所不同，例如：美國統一商法典第 2-201 條規定，除該法典另有規定外，貨物買賣之價金為五百元或五百元以上者，除非有書面文件為之證明當事人間確已成立契約，否則該契約不生效力（中興大學法律研究所：美國統一商法典及其譯注，第 43 頁以下）。美國統一商法典之規範，乃是源自於詐欺防止法 (Statute of Frauds)，為避免詐欺而規定買賣契約之形式要件。然於聯合國國際貨物買賣契約公約中，乃規範買賣契約無須一定格式，亦無須以書面或其他特定方式訂立，亦不須以任何書面作為證據，得以口頭或其他方式為之（買賣公約第一一條；劉春堂：論國際貨物買賣契約之締結，輔仁法學第九期，第 22 頁以下）。

　　就買賣價金與標的物之確定，乃屬標的物之適法、可能、確定之要素，故宜認為如對價金無從確定者，契約即為無效。然買賣公約 (CISG) 乃規定當事人雖未明示或默示價金額度或如何確定價金，然於訂約時存有慣常所要求之價金，且對契約之他造而言，此價金可確定者，即使當事人未明定價金，亦可認為默示同意出賣人之價金（買賣公約第一四條第一項）。再者，當事人未默示或明示規定價金或如何確定價金者，在沒有相反之意思表示下，視為雙方當事人默示訂約時此種有關貿易類似狀況下銷售之通常價金（買賣公約第五五條）。由此觀之，買賣公約就價金與我國民法第三四五條之規範不同，其得以訂約時通常之銷售價金定之（劉春堂：前揭文，第 7 頁以下）。由於買賣公約已為多數國家所接受，該立法之形式應值參考。

各國法規──Think as globally as possible.

Sales Agreement (Taiwan)

■ A Sale Agreement is a contract whereby the parties agree that one of them shall transfer to the other his rights over property and the latter shall pay a price for it.

The contract of sale is completed when the parties have mutually agreed on the object and the price. (345)

■ The provisions under the present title shall apply mutatis mutandis to such non-gratuitous contracts other than those of sale, unless the nature of the contract does not permit. (347)

　　茲將買賣之承諾方式，列舉實例如下：

This order is accepted subject to the terms and conditions appearing on the reverse side hereof.

§3 瑕疵擔保的義務 (Conformity of the Goods and the Third Party Claims)

出賣人就出賣之標的物負有瑕疵擔保之責任，該瑕疵擔保責任包括權利之瑕疵擔保（民法第三四九條以下），及物之瑕疵擔保（民法第三五四條以下）。以下乃就瑕疵擔保責任之性質、該兩種瑕疵擔保之要件及內容為介述：

出賣人之瑕疵擔保責任乃法定責任，即當事人於訂立買賣契約時，不必就是否有瑕疵擔保為特別約定或意思表示，出賣人一經成立買賣關係即負有瑕疵擔保責任，此即所謂之法定責任（鄭玉波：民法債編各論，第 29 頁以下），惟此等法定責任得由當事人約定予以限制、免除或加重（參考民法第三六六條）。

1.權利瑕疵擔保

民法第三四九條至第三五三條乃規定買賣之出賣人權利瑕疵擔保，該規定乃主要為權利瑕疵擔保之要件、當事人特約及法律效果，以下就前述事項分別說明之：

⑴要　件

民法第三四九條及第三五〇條乃規定權利瑕疵擔保之成立要件，以下乃就該條文之規範說明之：

①權利無缺之擔保：就出賣標的物而言，出賣人應擔保第三人對買受人不得主張任何權利（民法第三四九條）。

②權利存在之擔保：如出賣之標的物為權利時，出賣人應擔保權利確實存在。

③權利之瑕疵為自始不存在或欠缺：權利之瑕疵需為自始之主觀給付

不能，如為自始客觀不能，乃契約無效之問題（民法第二四六條、第二四七條）。

④買受人為善意：即買受人於契約成立時，需不知權利有所瑕疵（民法第三五一條）。

⑤權利瑕疵無從事後去除：依民法第三四八條之規定，權利之出賣人負有使買受人取得權利之義務（民法第三四八條第二項）。

⑵特　約

就權利之瑕疵擔保，當事人得以特約限制或免除此等責任，但為保障公序良俗，如其為出賣人所故意不告知者，其特約即為無效（民法第三六六條）。就各國之立法例而言，均贊成依私法自治原則，由當事人免除此種擔保責任（買賣公約第三五條第二項、第四一條第一項）。然英美法認為此種限制或免除，對買受人較為不利，故對有關擔保之免除或修正（Exclusion or Modification of Warranties），其若以書面為之者，其免除需為明顯之標示美國統一商法典（第 2–316 條第 2 項）。

⑶法律效果

就權利瑕疵擔保責任之法律效果而言，出賣人應依債務不履行之規定負其責任（民法第三五三條）。即需依分別有關給付遲延、給付不能及不完全給付之狀況，要求予以損害賠償、解除契約或要求違約金。

2.物之瑕疵擔保

⑴基本概念

出賣人對買受人之第二個責任，該責任乃指物之出賣人就物本身之瑕疵，應負擔保責任（鄭玉波：前揭書，第 41 頁）。

物之瑕疵擔保乃指物之買賣所生之擔保責任，對權利之出賣，如其亦須交付特定物者（民法第三四八條第二項），亦準用物之瑕疵擔保之規定（鄭玉波：前揭書，第 41 頁）。出賣人對於買受人所負物之瑕疵擔保責任，乃當然應負之法定責任，其係為補充當事人之意思而設，故即使當事人未明示負擔保責任，出賣人亦負有此種責任（八十二年臺上字第三二八八號

判決）。故英美法上，乃將當事人特約之瑕疵擔保責任，稱為明示之擔保責任 (Express warranty)，而當事人未有約定者，即形成默示之擔保責任 (Implied warranty)（美國統一商法典第 2–313 條）。

(2)**物之瑕疵擔保之內容**

依民法第三五四條第一項之規定，物之出賣人對於買受人應擔保其物依第三七三條之規定危險移轉於買受人時，無滅失或減少其價值之瑕疵，亦無滅失或減少其通常效用或契約預定效用之瑕疵，但減少之程度無關重要者不得視為瑕疵。再者，出賣人並應擔保其物於危險移轉時，具有所保證之品質。由此規定，可知物之瑕疵擔保所擔保之瑕疵包括：價值之瑕疵、預定或通常效用之瑕疵及保證品質所生之瑕疵（民法第三五四條第二項）。

①價值瑕疵之擔保

出賣人於危險移轉於買受人時，須擔保其物並無滅失或減少其價值之瑕疵，所謂之價值瑕疵擔保，乃指交換價值而言，使用價值乃預定或通常效用，並非此項所稱之價值瑕疵（鄭玉波：前揭書，第 41 頁），影響使用價值者，常致交換價值降低，而交換價值降低者，並不當然致使用價值減損（鄭玉波：前揭書，同頁）。

②效用瑕疵之擔保

物之出賣人就標的物無滅失或減少其通常效用或預定效用，所謂之通常效用，指一般效益上所應具有之效用，而所謂之預定效用，乃指該物在一般觀念上並無此效用，然當事人特以契約意定此效用（鄭玉波：前揭書，第 42 頁）。此於買賣公約即所謂之一般目的 (Ordinary Purpose) 及特別目的 (Particular Purpose)（買賣公約第三五條第二項）。此種特定效用，得為明示或默示之方法，而成為當事人合意預定之效用（買賣公約第三五條第二項）。如出賣人交付於買受人之特定樣品 (Sample, Model) 即構成貨樣買賣，則出賣人應擔保其交付之標的物，與貨樣有同樣之品質（民法第三八八條）。

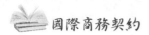

各國法規——Think as globally as possible.

Warranty: Conformity of the Goods and the Third Party Claims (Taiwan)

■ The seller shall warrant that the thing sold is free from any right enforceable by third parties against the buyer. (Taiwan Civil Code, 349) The seller of a claim of prestation or any other right shall warrant the actual existence of such prestation or right. The seller of valuable securities shall also warrant that it shall not be declared voidance through public summons. (350)

■ The seller of a thing shall warrant that the thing sold is, at the time when the danger passes to the buyer according to the provisions of Article 373, free from any defect in quality which may destroy or impair its value, or its fitness for ordinary efficacy, or its fitness for the efficacy of the contract of sale. However, if the extent of the impairment is of no importance, such impairment shall not be deemed to be a defect.

The seller also shall warrant that, at the time the danger passes; the thing has the guaranteed qualities. (354)

各國法規——Think as globally as possible.

Warranties (U.C.C.)

■ There is in a contract for sale a warranty by the seller that: (a) the title conveyed shall be good and its transfer rightful and shall not unreasonably expose the buyer to litigation because of any colorable claim to or interest in the goods; and (b) the goods shall be delivered free from any security interest or other lien or encumbrance of which the buyer at the time of contracting has no knowledge. (U.C.C. 2–312)

■ Express warranties by the seller to the immediate buyer are created as follows: (a) Any affirmation of fact or promise made by the seller which relates to the goods and becomes part of the basis of the bargain creates an express warranty that the goods shall conform to the affirmation or promise. (b) Any description of the goods which is made part of the basis of the bargain creates an express warranty that the goods shall conform to the description. (2-313)

■ Unless excluded or modified, a warranty that the goods shall be merchantable is implied in a contract for their sale if the seller is a merchant with respect to goods of that kind. (2-314)

茲將一般商品交易瑕疵擔保之約定條款及擔保排除之範例說明如下：

Seller warrants that the goods covered by this order conform to contract specifications. ALL OTHER WARRANTIES, EXPRESS OR IMPLIED, INCLUDING WITHOUT LIMITATION ANY IMPLIED WARRANTY OF MERCHANTABILITY, ARE EXCLUDED.

Seller warrants that goods are manufactured pursuant to specifications and have full merchantability, but makes no additional express or implied warranties or representations of any whatsoever with regard thereto.

然該約定應遵守各法律之強制規定。

各國法規——Think as globally as possible.

Exclusion or Modification of Warranties (U.C.C.)

■ Words or conduct relevant to the creation of an **express warranty** and words or conduct tending to negate or limit warranty shall be construed wherever reasonable as consistent with each other.

> ■ To exclude or modify the **implied warranty** of merchantability or any part of it in a consumer contract the language must be in a record, be conspicuous, and state "The seller undertakes no responsibility for the quality of the goods except as otherwise provided in this contract."
>
> ■ Language to exclude **all implied warranties** of fitness in a consumer contract must state "The seller assumes no responsibility that the goods will be fit for any particular purpose for which you may be buying these goods, except as otherwise provided in the contract." (U.C.C. 2–316)

⑶瑕疵擔保之要件

買受人就標的物之瑕疵符合下列之要件時，即可要求出賣人負擔物之瑕疵擔保責任：

①須標的物依民法第三七三條之規定於危險移轉於買受人時，該物具有瑕疵（民法第三五四條第一項前段）。

②買受人為善意且無重大過失。

③買受人負檢查及通知之義務。

⑷物之瑕疵擔保之效果

如出賣人之物之瑕疵擔保成立，買受人即得主張債務不履行或民法第三五九條以下之各項權利。該等權利包括解除契約、減少價金、請求不履行之損害賠償及另行交付無瑕疵之物。

①解除契約

買賣因物有瑕疵者，且出賣人應負瑕疵擔保之時，買受人得解除契約（民法第三五九條前段）。買受人固得主張解除契約，出賣人於買受人主張物有瑕疵者，亦得訂期限催告其是否解除契約，買受人於一定期限內不為解除契約者，喪失其解除契約之權利（民法第三六一條）。然喪失其解除權，並不當然喪失價金減少或不履行時損害賠償之權利。

②減少價金

出賣人應負瑕疵擔保責任時，買受人得解除契約或請求減少價金，即

為兩項權利乃並行，而由買受人選擇行使之。但依其情形，解除契約顯失公平者，買受人僅得請求減少價金（民法第三五九條）。

③損害賠償請求權

如出賣人保證標的物之品質，而該標的物無此等之品質時，買受人尚得請求不履行之損害賠償，或出賣人故意不告知物之瑕疵者亦同。

④另行交付請求權

買賣之標的物如僅為指定種類者，如其物有瑕疵時，買受人亦得不解除契約或請求減少價金，而請求另行交付無瑕疵之物（民法第三六四條第一項），就另行交付之物，出賣人仍負擔保責任（民法第三六四條第二項）。

雖有前述之救濟方式，當事人亦得限制賠償之額度，茲就賣方對瑕疵擔保責任之限制約定條款，舉例說明如下：

If not replaced by seller as herein provided, seller's liability shall be limited to the stated selling price of any defective goods. Seller shall in no event be liable for Buyer's manufacturing costs, lost profits, good will or other special or consequential damages.

---

(5)物之瑕疵擔保之特約

就物之瑕疵擔保責任，當事人得為加重或減免其責任之特約：

①依民法第三六六條之規定，買賣之當事人得以特約免除或限制出賣人關於物之瑕疵擔保之義務，但如出賣人故意不告知其瑕疵者，其特約無效。

②加重責任之特約包括無瑕疵之特約與保證品質之特約，就保證無瑕疵之特約而言，其效果規定於民法第三五五條第二項，即如出賣人保證無瑕疵者，縱買受人有重大過失而不知其瑕疵者，出賣人仍應負瑕疵擔保責任。

③依民法第三六五條第一項之規定，就解除契約及減價之除斥期間為買受人依第三五六條為通知後六個月，或自物之交付時起經過五年，但當事人得另為約定並延長期間，以符合契約自由原則。

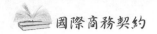

§4 價金支付的義務 (Payment of the Price)

依民法第三六七條之規定，買受人對出賣人有支付約定價金之義務。以下乃就價金之支付數額、處所、時期及拒絕為說明。

1. 價金之計算

按價金之數額乃依當事人之約定定之，當事人如未為約定者，而依客觀情形可得而定者，乃依情形所定之價金為準（民法第三四六條第一項）。再者，如約定依市價者，應以清償時清償地之市價為準（民法第三四六條第二項）。價金依物之重量而計算者，應除去包皮之重量。但契約另有訂定或另有習慣者，應依該契約之約定或習慣（民法第三七二條）。

2. 交付之處所

就價金交付之處所，應依當事人之約定為之，惟標的物與價金應同時交付者，價金應於標的物之交付處所交付之（民法第三七一條）。

3. 交付物之時期

除法律、契約另有規定或另有習慣外，標的物與價金交付應同時為之（民法第三六九條）。即標的物訂有期限者，其價金亦應依期限為之交付（民法第三七〇條）。依此而論，標的物交付後，買受人即應交付價金，於期限屆至前買受人無須支付利息，於期限屆滿後，買受人即應支付遲延利息。

4. 給付之拒絕

如買受人有正當理由，認第三人對標的物將主張權利，而使其買賣契約所生之權利將喪失者，得拒絕交付價金之全部或一部，惟出賣人已提供擔保，或提存價金者，買受人即不得拒絕支付價金（民法第三六八條）。

§5 受領義務 (Taking Delivery)

依民法第三六七條之規定，買受人對標的物有受領之義務，如其受領遲延時，即構成債權人之債務不履行，應依債務不履行之規定負起責任。買受人受領標的物之義務，應以出賣人依債務本旨提出給付為要件，如出賣人交付之物有瑕疵者，買受人仍得拒絕受領（黃茂榮：前揭書，第 516 頁）。然如為異地買賣者，買受人負有保管之責，並於一定情況下，負有變賣及通知之義務，否則，買受人應負損害賠償之責（民法第三五八條）。

如買受人未盡受領標的物之義務時，買受人應負受領遲延及給付遲延之責任，因該規定之受領，亦為一種給付（鄭玉波：前揭書，第 68 頁）：

1.受領遲延

於出賣人受領遲延時，出賣人僅就其故意或重大過失負起責任（民法第二三七條），並得要求買受人負擔請求賠償及保管給付物之必要費用（民法第二四〇條）。再者，出賣人亦得拋棄占有或提存、拍賣標的物，而提存價金（民法第二四一條、第三二六條）。

2.給付遲延

其出賣人得請求因不受領所生之損害賠償（民法第二三一條、第二三二條），及解除契約（民法第二五四條）。

就買受人之義務而言，現行法之規定，與買賣公約之規範相同（買賣公約第五三條）。惟就價金之支付時期、地點等，我國民法與買賣公約有諸多不同之處（買賣公約第五七條以下）。

§6　利益之承受及危險負擔 (Passing of Risk, Risk of Loss)

　　按買賣標的物之利益及危險，自交付時起，均由買受人承受負擔。此為我國民法對利益承受及危險負擔之基本規定，以下就其相關之問題說明如下：

1.基本原則

　　依民法第三七三條之規定，就利益承受與危險負擔之決定始點，如標的為物者，應以物之交付為始點，以權利為標的者，亦以此為始點（民法第三七七條）。此所稱之交付包括現實交付與觀念交付，所謂之觀念交付包括：簡易交付、占有改定及指示交付（民法第七六一條）。

2.例外之情形

　　不論利益之承受或危險之負擔，如當事人另有約定者，或法律另有規定時，均依該約定或法律規定為決定其利益承受與危險負擔之始點（鄭玉波：前揭書，第71頁以下）。如買受人請求出賣人將標的物送交清償地以外之處所者，自出賣人交付其標的物於運送承攬人時起，買受人即負擔標的物之危險（民法第三七四條）。但如契約當事人約定，出賣人負有義務將標的物送交某一處所者，於該標的物到達該地時，其利益與危險由買受人承受（比較買賣公約第六七條）。

各國法規——Think as globally as possible.

Passing of Risk (Taiwan)

■ The profits and dangers of the object sold pass to the buyer at the time of delivery, unless otherwise provided by contract. (Taiwan Civil Code, 373)

■ If the buyer requests that the object sold be delivered at a place other than the place where delivery ought to be made, the dangers pass to the buyer at the time when the seller delivers the object to the person who transports it or is entrusted with its transportation. (374)

■ When the object of a sale is a right, by virtue of which the seller may possess a certain thing, the provisions of the preceding articles shall be mutatis mutandis applied. (377)

各國法規──Think as globally as possible.

Risk of Loss (U.C.C.)

■ (a) if it does not require him to deliver them at a particular destination, the risk of loss passes to the buyer when the goods are duly delivered to the carrier; but

■ (b) if it does require him to deliver them at a particular destination and the goods are there duly tendered while in the possession of the carrier, the risk of loss passes to the buyer when the goods are there duly so tendered as to enable the buyer to take delivery.

■ (1) Where a tender or delivery of goods so fails to conform to the contract as to give a right of rejection the risk of their loss remains on the seller until cure or acceptance.

■ (2) Where the buyer rightfully revokes acceptance he may to the extent of any deficiency in his effective insurance coverage treat the risk of loss as having rested on the seller from the beginning. (U.C.C. 2–509)

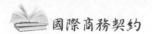

各國法規──Think as globally as possible.

Passing of Risk (CISG)

■ (1) If the contract of sale involves carriage of the goods and the seller is not bound to hand them over at a particular place, the risk passes to the buyer when the goods are handed over to the first carrier for transmission to the buyer in accordance with the contract of sale. If the seller is bound to hand the goods over to a carrier at a particular place, the risk does not pass to the buyer until the goods are handed over to the carrier at that place. The fact that the seller is authorized to retain documents controlling the disposition of the goods does not affect the passage of the risk. (CISG, 67–(1))

■ (2) Nevertheless, the risk does not pass to the buyer until the goods are clearly identified to the contract, whether by markings on the goods, by shipping documents, by notice given to the buyer or otherwise. (67–(2))

■ Loss of or damage to the goods after the risk has passed to the buyer does not discharge him from his obligation to pay the price, unless the loss or damage is due to an act or omission of the seller. (66)

§7　救　濟

買賣之當事人，得向對方為下述之求償：

(1)買賣之出賣人不履行權利瑕疵擔保者（民法第三四八－三五一條），買受人得依債務不履行之規定，即前章所述之給付不能、給付遲延、不完全給付等，行使買受人之權利（第三五三條）。

(2)買賣因物有瑕疵，而出賣人應負物之瑕疵擔保責任者，買受人得解除其契約，或請求減少其價金。但依其情形，解除契約顯失公平者，買受人僅得減少價金（第三五九條）。

(3)買賣之物，缺少出賣人所保證之品質者，買受人得不解除契約或請求減少價金，而請求不履行之損害賠償。出賣人故意不告知物之瑕疵者，亦同（第三六〇條）。

美國 U.C.C. §2–703 及 §2–711 則分別規定賣方及買方請求救濟之方式及範圍。

各國法規──Think as globally as possible.

Remedies (Taiwan)

■ If the seller does not perform his duties specified in Articles 348–351, the buyer may exercise his rights in accordance with the provisions concerning non-performance of obligations. (Taiwan Civil Code, 353)

■ When there is a defect in the thing sold for which, according to the provisions of the five preceding articles, the seller is responsible for a warranty, the buyer has the option to rescind the contract or to ask for a reduction of the price, unless in the case specified, that a rescission of the contract would constitute an obvious unfairness of the transaction the buyer is only entitled to ask for a reduction of the price. (359)

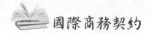

■ In the absence of a quality of the thing sold, which was guaranteed by the seller, the buyer may demand to compensate for the injury of non-performance, instead of rescission of the contract or of a reduction of the price. The same rule shall be applied if the seller has intentionally concealed a defect in a thing. (360)

各國法規──Think as globally as possible.

Remedies in General (CISG)

■ **Anticipatory breach** (suspend the performance (71(1)), prevent the handing over of the goods (CISG, 71(2))

■ **Fundamental breach of contract** (declare the contract avoided (72(1))─ Avoidance of the contract releases both parties from their obligations under it, subject to any damages which may be due.)

■ **Damages** (a sum equal to the loss, including loss of profit, suffered by the other party as a consequence of the breach (74), recover the difference between the contract price and the price in the substitute transaction (75), interests (78).

各國法規──Think as globally as possible.

Seller's Remedies (CISG)

■ (1) If the buyer fails to perform any of his obligations under the contract or this Convention, the seller may:

(a) exercise the rights provided in Articles 62 to 65 (require to perform obligations, declare the contract avoided, etc.);

(b) claim damages as provided in Articles 74 to 77 (suspend the performance.)

(2) The seller is not deprived of any right he may have to claim damages

by exercising his right to other remedies.

(3) No period of grace may be granted to the buyer by a court or arbitral tribunal when the seller resorts to a remedy for breach of contract.

各國法規——Think as globally as possible.

Seller's Remedies (U.C.C.)

■ (1) A breach of contract by the buyer includes the buyer's wrongful rejection or wrongful attempt to revoke acceptance of goods, wrongful failure to perform a contractual obligation, failure to make a payment when due, and repudiation.

■ (2) If the buyer is in breach of contract the seller, to the extent provided for by this Act or other law, may:

(a) withhold delivery of such goods;

(b) stop delivery of the goods under Section 2–705;

(c) proceed under Section 2–704 with respect to goods unidentified to the contract or unfinished;

(d) reclaim the goods under Section 2–507 (2) or 2–702 (2);

(e) require payment directly from the buyer under Section 2–325 (c);

(f) cancel;

(g) resell and recover damages under Section 2–706;

(h) recover damages for non-acceptance or repudiation under Section 2–708 (1) or in a proper case the price (Section 2–709);

(j) recover the price under Section 2–709;

(k) obtain specific performance under Section 2–716;

(l) recover liquidated damages under Section 2–718;

(m) in other cases, recover damages in any manner that is reasonable under the circumstances. (U.C.C. 2–703)

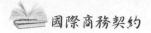

Buyer's Remedies (U.C.C.)

■(1) A breach of contract by the seller includes the seller's wrongful failure to deliver or to perform a contractual obligation, making of a nonconforming tender of delivery or performance, and repudiation.

■(2) If the seller is in breach of contract under subsection (1), the buyer, to the extent provided for by this Act or other law, may:

(a) in the case of rightful cancellation, rightful rejection, or justifiable revocation of acceptance, recover so much of the price as has been paid;

(b) deduct damages from any part of the price still due under Section 2–717;

(c) cancel;

(d) cover and have damages under Section 2–712 as to all goods affected whether or not they have been identified to the contract;

(e) recover damages for nondelivery or repudiation under Section 2–713;

(f) recover damages for breach with regard to accepted goods or breach with regard to a remedial promise under Section 2–714;

(g) recover identified goods under Section 2–502;

(h) obtain specific performance or obtain the goods by replevin or similar remedy under Section 2–716;

(i) recover liquidated damages under Section 2–718;

(j) in other cases, recover damages in any manner that is reasonable under the circumstances. (U.C.C. 2–711).

第五篇

各種契約範例（中英對照）

§1 買賣契約 (Sale)

所謂買賣 (Sale) 乃指當事人約定由出賣人 (Seller) 移轉財產權於買受人 (Buyer)，買受人支付價金之契約。買賣契約一般應包括下列條款：

1. 訂約日期 (Date of Agreement)。
2. 當事人姓名及地址 (Names and addresses of parties)。
3. 出賣人之出售與交付義務 (Duties of seller to sell and deliver)。
4. 買受人之購買與受領義務 (Duties of purchase and accept delivery)。
5. 銷售之標的 (Description of goods and chattel)。
6. 數量 (Quantity)。
7. 價格 (Price)。
8. 價金之給付 (Payment)。
9. 物之交付 (Delivery of goods)。
10. 標的之檢查及受領 (Inspection and acceptance of goods)。
11. 待同意之出售 (Sale on Approval)。
12. 擔保 (Warranties)。
13. 契約違反之救濟 (Remedies for breach)。
14. 不履行之免責 (Waiver of nonperformance)。
15. 契約之修正與終止 (Modification and cancellation of contract)。
16. 契約權義之轉讓 (Assignment of contract rights and delegation of duties)。
17. 銷售稅捐 (Sales taxes)。
18. 契約之期間 (Term of agreement)。
19. 契約之更新 (Renewal of agreement)。
20. 糾紛之解決 (Dispute settlement)。
21. 準據法 (Governing law)。

CONTRACT

1. Obj... ...his Contract

1.1. ...tomer shall order and the Executor shal'...

consu... ...der the Technical Assignment (A...

Contra... ...ges) of the performance of...

1.2. Per... ...nment.

Technica...

2. Obligatio... ...e Parties

2.1. The Cu... ...T shall be obliged:

a) to p... ...he work perform...

Contrac... ...mely all nece...

b) to pro... ...o provide f...

Executor; ...be oblige...

c) if necess... ...under th...

2.2. The Executo... ...e of wor...

a) to perform... ...h the ful...

b) upon perfo... ...ntract a...

c) to perform v... ...ntract...

determined in th... ...m work t...

3. Procedure of Work ...any third f...

3.1. The Executor shall p... ...conditions o...

3.2. The Executor may e... ...pproval or not...

however, subject to terms ...

...led for in this Contract...

...t required. ...ion of work and within the period in...

...e the Report on the performed work and...

...ls the Customer for signing.

...d days of the date of receipt of the Report and

...ction. the performed work shall be dee...

...t shall be argumented and inclu...

...assignment. In such...

...ion.

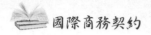

出賣確認書

××公司

售予：A 公司地址： ＿＿＿＿＿＿＿＿＿＿＿＿＿

　　　　日期： ＿＿＿＿＿＿＿＿＿＿＿＿＿＿＿

＿＿＿＿＿＿本公司檔案號碼： ＿＿＿＿＿＿＿＿＿

＿＿＿＿＿＿客戶訂單號碼： ＿＿＿＿＿＿＿＿＿＿

我們以下乃感謝收到台端之訂單如下屬：

（訂單之內容）

台端之要約乃依本確認書背面所載之條款而經接受，該條款不得予以變更，請於確認書簽名並將副本予以返還。

B 公司

＿＿＿＿＿＿＿＿＿＿

（賣方簽名）

確認：

＿＿＿＿＿＿＿＿＿＿

（買方簽名）

SELLER'S ORDER CONFIRMATION

×× CORPORATION

SOLD TO: A Company Address: _____

Date: _____

_____ Our Ref. No.: _____

_____ Cust. Order No.: _____

We hereby acknowledge with thanks, receipt of your order as set forth below:

[fill in details]

THIS ORDER IS ACCEPTED SUBJECT TO THE TERMS AND CONDITIONS APPEARING ON THE REVERSE SIDE HEREOF. NO CHANGES MAY BE MADE THEREIN. PLEASE SIGN AND PROMPTLY RETURN THE DUPLICATE COPY.

B CORPORATION

By _____

Signature

ACCEPTED:

_____

Buyer's Signature

出賣之條款

第一條　台端之要約乃依以下之條件而為接受，該條件非經賣方合法之授權代表人之書面承認，否則，不得予以變更。

第二條　除本確認書另有記載者外，賣方就運費、運送費用、保險費、海運費、倉儲費、處理費、滯留費及其他類似之費用不負其責任。如前述之費用已包括於售價中者，於本確認書記載之日期以後所生費用之增加，由買方負擔之。

第三條　除本確認書另有記載者外，所有之進口許可及執照，及有關美國輸入關稅或關稅手續費之費用，均由買方負擔之。

第四條　除法律另有規定者外，所有之銷售稅或其他類似之稅捐，應等確認書出售物品所須負擔或徵收之稅捐，均由買方負擔之。

第五條　賣方就因非經賣方所得以控制之事由所生之交付遲延或不能交付不負其責任，該事由包括但不限於：自然行為、戰爭、動員、騷亂、暴亂、禁運、本國或外國政府之法規、火災、水災、罷工、停工及其他勞工之爭議、或其他應欠缺或無法獲得運輸工具者。

第六條　買方非經書面逾十日之通知，不得以交付之遲延請求撤銷本確認書，且於貨物運送之後，不得以任何之事由解除其購買之通知。

第七條　超出本確認書表面記載之說明之範圍者，賣方不負擔任何明示或默示有關物品商品性之擔保責任。

TERMS AND CONDITIONS OF SALE

Section 1. This order is accepted on and subject to the following terms and conditions, which may not be modified except by writing signed by Seller's duly authorized representative.

Section 2. Except as otherwise provided herein, Seller shall not be responsible for freight, transportation, insurance, shipping, storage, handling, demurrage or similar charges. If such charges are by the terms of sale included in the price, any increase in rates becoming effective after the date hereof shall be for the account of the Buyer.

Section 3. Except as otherwise provided herein, all import permits and licenses and the payment of all United States import duties and customs fees shall be sole responsibility of the Buyer.

Section 4. All sales and similar taxes which the Seller may be required to pay or collect with respect to the goods covered by this order shall be for the account of the Buyer, except as otherwise provided by law.

Section 5. Seller shall not be responsible for delays in delivery or and failure to deliver due to causes beyond Seller's control, including but not limited to acts of God, war, mobilization, civil commotion, riots, embargoes, domestic or foreign governmental regulations or orders, fires, floods, strikes, lockouts and other labor difficulties, or shortages of or inability to obtain transportation.

Section 6. Buyer may not in any event cancel this order for any delays in delivery without giving at least ten days prior written notice of intention to do so, and in no event after goods have left point of shipment.

Section 7. There are no warranties, express or implied including merchantability, which entered beyond the description on the face

第八條　賣方就物品有關耐度、容積、重量及數量之通常偏差不負其責任。物品之重量、大小及數量乃由賣方之工廠或其他機關為終局之決定。

第九條　除本確認書另有記載者外，CIF 買賣保險乃依 F.P.A. 之條件。

第十條　如當事人就物品之檢查或測試已同意者，該測試或檢查須於賣方之工廠或其他之機關於運送前為之，買方就物品之承認或拒絕須於運送前為之。於運送之後不得為異議。就其他之情形，買方須於受領物品十五日內且有合理之機會檢查該物品之狀況下，以書面向賣方提出請求。

第十一條　如出售予買方之物品有瑕疵或與契約所訂之規格不合者，賣方有權依其裁量將物品為替換或將買得之價金依其部分予以退還，非經賣方之書面同意，不得將物品退還於買方。於任何狀況下，賣方不就買方有關製造費用、損失之利益、信用之損害或其他特別或結果之損害負其責任。

第十二條　除本確認書另有記載者外，賣方有權分批交付其貨品，該分期之交付須有分別之收據，買方且得不論其後續送達之物品而得以分批給付價金。如買方未能於到期時分批給付價款，賣方得就後續之交付免除其責任。單一交付之遲延，不得免除買方接

hereof.

Section 8. Seller shall not be liable for normal variations in tolerance, dimensions, weights and quantity. Weights, sizes and quantities as determined at Seller's mill or other source of supply shall be conclusive.

Section 9. Except as otherwise provided, insurance on CIF sales shall be F.P.A. terms.

Section 10. If an inspection or testing of the goods has been agreed upon, such inspection or testing shall be made at Seller's mill or other source of supply before shipment of the goods, and approval or rejection shall be made promptly and in any event before shipment. No claims will be entertained thereafter. In all other cases Buyer is required to give written notice to Seller of any claim promptly upon receipt of the goods and in any event within fifteen (15) days thereafter, and Seller shall thereupon be afforded a reasonable opportunity to inspect the goods.

Section 11. If any portion of the goods delivered to Buyer are defective or are otherwise not in accordance with contract specifications, Seller shall have the right in its discretion either to replace such defective goods or to refund the portion of the purchase price applicable thereto. No goods shall be returned to Seller without Seller's written consent. In no event shall Seller be liable for the cost of processing, lost profits, injury to good will or any other special or consequential damage.

Section 12. Unless otherwise expressly stated, Seller shall have the right to make delivery in installments. All installments shall be separately invoiced and paid as billed without regard to subsequent deliveries. Failure to pay for any installment when due shall excuse Seller

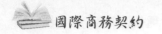

受後續交付之責任。

第十三條　如賣方就買賣予以買方信用者，且買方不能支付前已支付之價
　　　　　金或依賣方之意見，買方之財務條件有重大變更時，賣方得取
　　　　　消其給予買方之信用，賣方且於後續運送前，要求買方為支付。

第十四條　任何有關本契約之爭議或請求或有關本契約違反之事項，當事
　　　　　人須依美國仲裁協會之規則與紐約仲裁之方式解決之，就仲裁
　　　　　之判斷得經任何有管轄權之法院為認可。就仲裁之請求須於貨
　　　　　品交付後九十日內為請求。

第十五條　本契約不得以口頭之方式為修正或終止。非經賣方合法授權之
　　　　　代表出面簽署者，有關請求之變更、終了或權利之放棄不得生
　　　　　效。

第十六條　本契約以紐約州法為準據法且依該法而為解釋之。

from making further deliveries. Delay in delivery of any installment shall not relieve Buyer of its obligation to accept remaining installments.

Section 13. If by the terms of sale, credit is extended to Buyer, Seller reserves the right to revoke such credit if Buyer fails to pay for any goods previously delivered when due or if in the judgment of Seller there has been a material adverse change in Buyer's financial condition, and thereupon Seller have the right to demand payment before further shipment of any goods.

Section 14. All controversies and claims arising out of or relating to this contract, or the breach thereof, shall be settled solely by arbitration held in New York City in accordance with the rules then obtaining of the American Arbitration Association, and judgment upon any award thereon may be entered in any court having jurisdiction thereof. Any demand for arbitration hereunder shall be made not later than ninety (90) days after delivery of the goods.

Section 15. This contract may not be modified or terminated orally. No claimed modification, termination or waiver of any of its provisions shall be valid unless in writing signed by Seller's duly authorized representative.

Section 16. This contract shall be governed by and construed according to the laws of the State of New York.

§2 特定商品購買貨物契約

　　按前述買賣契約須記載之內容應包括：當事人、出賣人之義務、買受人之義務、標的、數量價格等事項外，就特定之標的物，當事人必須約定相關規格，以下之範例即為特定商品買賣貨物契約之範例，通常應包括下列條款：

　　1.價格 (Price)。

　　2.價金之給付 (Payment)。

　　3.貨品之規格 (Specifications)。

　　4.貨物製造及交付之完成日期 (Completion Schedule)。

　　5.物品交付之時間與地點 (Delivery)。

　　6.擔保 (Warranty)。

　　7.契約之終止 (Termination)。

　　8.得當事人之通知 (Notice)。

　　9.準據法及語言 (Governing Law and Language)。

CONTRACT

1. Obj ... **this Contract**
1.1. ... tomer shall order and the Executor shall ...
consu... der the Technical Assignment (Ap-
Contra ... ges) of the performance of ...
1.2. Per ... nment.
Technica...

2. Obligatio ... **e Parties**
2.1. The Cu... shall be obliged:
a) to pa... he work perform...
Contract ... mely all nec...
b) to pro... provide ...
Executor; ... be obliged ...
c) if necess... under th...
2.2. The Executo... e of wor...
a) to perform ... n the ful...
b) upon perfo... ntract a...
c) to perform ... m work ...
determined in th... any third ...

3. Procedure of Work ... conditions o...
3.1. The Executor shall ... pproval or not...
3.2. The Executor may e... subject to terms ...
... vided for in this Contract ...
... not required. ... letion of work and within the period ...
... the Report on the performed work an...
... the Report on the performed work and ...
... the Customer for signing. ... date of receipt of the Report an...
... y days of the date of receipt ... performed work shall be de...
... ection, the performed work shall be ...
... r shall be argumented and inclu...
... he assignment. In such...

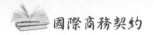

特定商品購買貨物契約

本契約於＿＿＿年＿＿＿月＿＿＿日由以下當事人簽訂之：

A，住所位於＿＿＿；及 B 公司，依中華民國法律設立，及其主事務所為＿＿＿＿＿＿＿＿。

B 公司為製造該貨物之製造商，A 預期 B 公司得製造特定商品，而 A 期待購買該貨品（「商品」），且 B 公司願意製造該商品且出售予 A。

依此，當事人合意如後：

第一條　價格

　　　1.該商品之價格為＿＿＿美金 (US$＿＿＿)。

　　　2.該價格之單位以美金為基礎，且包括第三條之特殊規格。

　　　3.購買該商品之價金非以書面之方式不得為增加或降低。

第二條　給付

　　　1.A 以信用狀方式支付＿＿＿% 為保證金，且於簽約後＿＿＿日內給付之。

　　　2.於 B 公司製作完成該商品，且出示得為檢查之證件以供交付時，該保證金應由 A 公司以不得撤回之信用狀行使支付予 B 公司。如 B 未能於＿＿＿年＿＿＿月＿＿＿日出示該檢查之證明，該等信用狀將不生效力，而 A 公司無義務給付該保證金。

CONTRACT FOR PURCHASE OF GOODS

THIS AGREEMENT made on the _____ day of _____ , 20 _____ by and between:

A, residing at _____ (hereinafter "A"); and B Inc., a company duly organized and existing under the laws of the Republic of China, with its principal place of business at _____ (hereinafter "B").

WHEREAS, B is in the business of manufacturing goods for private use, and

WHEREAS, A desires B to manufacture _____ (hereinafter "Goods") to his specifications which he desires to purchase, and B desires to manufacture and sell the Goods to A.

NOW THEREFORE, the parties agree as follows:

ARTICLE 1－PRICE

1. The purchase price of the Goods shall be _____ U.S. Dollars (US$ _____).
2. The price shall be fixed in U.S. Dollars and shall include the Goods with all specifications referred to in Article 3 hereof.
3. The purchase price specified in this Article shall not be increased or reduced except by written agreement signed by both parties.

ARTICLE 2－PAYMENT

1. A shall put up a deposit of _____ % of the purchase price (hereinafter "Deposit") within _____ (_____) days of execution of this contract by opening a letter of credit as indicated in Article 2, subclause 2.
2. The Deposit shall be in the form of an irrevocable letter of credit payable to B upon production by B of a certificate of inspection indicating that the goods has been completed according to the Specifications, as defined herein, and is ready for delivery. If B does not produce such certificate on or before _____ 20 _____ , such letter of credit shall be of no effect and A shall not be

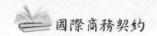

3.該超過保證金之剩餘部分應依第五條之規定給付予 B 公司。

4.前述之給付以美元計價之。

第三條　規格

1.就本商品之規格以附件 A 表示之，該商品之設計藍圖於附件 B 說明之，附件 A 及 B 為本契約之附件，此等規格、設計、設備與材料統稱為規格，B 公司必須嚴格遵守該等規格為製造。

2.規格所記載之項目以貿易使用之慣例定義之，如無貿易慣例者，以一般之英文解釋之。

3.本契約之規格應符合該商品之一般用語。

4.當事人得以書面之方式變更該規格。

第四條　完成時期

1.B 同意本商品之準時交貨為本契約必要且重要之點。

2.B 公司同意於 ＿＿＿ 年 ＿＿＿ 月 ＿＿＿ 日之前完成該商品且準備為交貨。

3.如該商品無法於 ＿＿＿ 年 ＿＿＿ 月 ＿＿＿ 日之前符合該規格者，B 公司同意於 ＿＿＿ 年 ＿＿＿ 月 ＿＿＿ 日之後，A 得單方決定終止本契約，而無須 B 公司之同意、法院之核准或仲裁，A 對 B 公司之義務乃依信用狀決定之。

obligated to pay the Deposit.

3. The balance of the purchase price in excess of the Deposit shall be payable to B upon delivery of the Goods according to Article 5 hereof.

4. Payment shall be calculated in U.S. Dollars.

ARTICLE 3—SPECIFICATIONS

1. The specifications for the Goods are set out in Attachment A. The design drawings of the Goods are specified in Attachment B. Attachments A and B are incorporated by reference herein. Such specifications and such design drawings, and equipment and materials specified therein, are collectively herein referred to as the "Specifications". B shall manufacture the Goods in strict accordance with the Specifications.

2. All items, in the Specifications shall be defined according to their usage in the trade or, if there is no established trade usage, according to their common English language meanings.

3. All Specifications shall be suitable and proper to all common usages of the Goods.

4. The Specifications shall not be amended or altered except by written agreement signed by both parties.

ARTICLE 4—COMPLETION SCHEDULE

1. B agrees that timely delivery is the essential and most important point of this Agreement.

2. B agrees to complete manufacture of the Goods and to be prepared to deliver it no later than _____ , 20 _____ .

3. If the Goods are not fully completed strictly according to the Specifications and ready for delivery on or before _____ , 20 _____ . B agrees that, at any time after _____ , 20 _____ and before accepting delivery of the Goods. A has the right, by his sole decision without B's further agreement and without the approval or authorization of any court, agency or arbitrator, to terminate

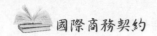

4.B 公司同意 A 或其代理人得隨時檢查該商品，以便於該商品之
製造或使其符合規格之變更者。

第五條　交付

1.本商品以＿＿＿為交貨地點並依 A 或其代理人之指示於＿＿＿
年＿＿＿月＿＿＿日為交付。

2.B 應提供設施以供商品之運送，或依運送之廠商及保險公司之
規定準備商品供遠洋之運送。

3.B 公司就商品到達＿＿＿之前之保險與損害負其責任。

第六條　擔保

B 公司同意以中華民國同產業之擔保為範圍，擔保商品之隱藏或
潛在的瑕疵。

第七條　終止

1.本契約得由當事人合意終止之。

2.本契約於當事人交付商品或給付價金或終止之。

3.本契約得依前述 4.3 條終止之。

4.於契約終止後當事人仍應就所同意繼續有效或實施之部分，就
該部分仍為有效。

第八條　通知

就當事人依本契約之通知，應以掛號函送前述之地址，或以傳真
及掛號信函送達前述之地址。該通知於接受前述掛號或傳真之方

this Agreement and all A's obligations to B under this Agreement including all obligations pursuant to any letters of credit.

4. B agrees that A or his agent may inspect the Goods at any time while it is being manufactured and require alterations to conform to the Specifications.

ARTICLE 5－DELIVERY

1. The Goods shall be delivered to ship's side at ＿＿＿ on or after ＿＿＿ , 20 ＿＿ in accordance with the instructions of A or his agent.

2. B shall provide a shipping cradle suitable for shipment of the Goods and otherwise prepare the Goods for overseas shipment in accordance with the requirements of the shipping line and insurance company selected by A.

3. B shall be responsible for insurance and loss until delivery is accepted at ship's side in ＿＿＿ .

ARTICLE 6－WARRANTY

B shall provide warranties against latent defects which are comparable to warranties provided by other ROC manufacturers in the industry.

ARTICLE 7－TERMINATION

1. This Agreement may be terminated at any time by mutual agreement of the parties.

2. This Agreement shall automatically terminate upon delivery of the Goods and payment of the purchase price in accordance herewith.

3. This Agreement may be terminated by A according to Article 4, subclause 3 hereof.

4. Following termination, the parties shall still be subject to the provisions of this Agreement as are intended to survive, operate or have effect after such termination.

ARTICLE 8－NOTICE

Any notice required or permitted under this Agreement may be made by registered mail to the respective addresses of the parties first above written, or

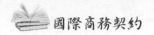

式之日起生效。

第九條　一般條款

1.當事人任一方為主張嚴格履行本契約條款時，該等主張不構成當事人契約權利之放棄。

2.本契約關於商品之條款優先適用於其他之前協議，不論其為口頭或書面之協議。

3.非經當事人以書面且簽署之方式為之，本契約之任何變更不生效力。

4.本契約以中華民國法律為準據法。

5.本契約以英文簽署之，B 公司同意其代表人瞭解本契約之條文，如本契約之中文與英文有相衝突者，本英文契約適用之。

基於前述內容，當事人於前述日期簽署本契約。

A　　　　　　　　　　　　B 公司

_____　　　_____

by fax, followed by registered mail to such addresses. The effective date of notice shall be the date of receipt of the registered mail if no other means of notice is used.

ARTICLE 9－GENERAL PROVISIONS

1. Failure of a party to insist on the strict performance of any provision shall not constitute waiver of such party's right to require such performance or later performance of a similar nature.

2. This Agreement supersedes all previous agreements, oral or written, with respect to the Goods.

3. No change, alteration or modification to this Agreement will be effective unless made in writing and signed by both parties.

4. This Agreement shall be governed by the laws of the Republic of China.

5. This Agreement has been negotiated and signed in the English language. B acknowledges that its representative has read and understood this Agreement. In the event of a conflict between this Agreement and a Chinese language translation, this Agreement shall govern.

IN WITNESS WHEREOF, the parties have executed this Agreement on the date first above written.

A   B

By _____   By _____

§3 經銷商品契約 (Sales Distribution Agreement)

經銷商品契約依其性質可為委任、行紀、居間、代辦商或買賣契約，其一般均包括下列條款：

1. 訂約之日期 (Date of agreement)。
2. 當事人之姓名及地址 (Names and addresses of parties)。
3. 貨物之銷售或銷售權之賦予 (Sale of goods or grant of selling privileges)。
4. 貨物之範圍 (Products or goods covered)。
5. 銷售之地區 (Territory covered)。
6. 貨物之買受價格 (Purchase price of goods)。
7. 經銷商同意之條款 (Acceptance of provisions by distributor)。
8. 最低購買數量 (Minimum purchase to be made)。
9. 經銷商非製造商之代理人 (Distributor not agent or legal representative of manufacturer)。
10. 契約期間 (Term of agreement)。
11. 契約之終止 (Cancellation or termination of agreement)。
12. 型式不得變更 (Forms not to be changed)。
13. 契約之排他性 (Exclusive nature of agreement)。
14. 公司保留之權利 (Rights reserved by company)。
15. 禁止銷售境外之義務 (Distributor not to sell outside of territory)。
16. 經銷商之給付 (Payment by distributor)。
17. 訂單之接受 (Acceptance of orders)。
18. 服務之維持 (Service installation and maintenance)。
19. 貨物或零件之返還 (Return of products or parts)。
20. 貨物之運送 (Shipment of product)。

21.帳冊之查核 (Auditing of books)。

22.運送人乃經銷商之代理人 (Common carriers as agents of distributor)。

23.經銷商之報告 (Reports by distributor)。

24.公司買回之權利 (Company's right to repurchase)。

25. 運送及安裝費用之收取 (Collection of freight and installation charges)。

26.使用製造商名稱之權利 (Right to use manufacturer's trade name)。

27.於境外銷售之調整 (Adjustment of sales made in other territory)。

28.對政府之銷售 (Sales to government)。

29.讓與契約之禁止 (Assignment of agreement prohibited)。

30.展示汽車或設備 (Demonstrating car or equipment)。

31.終止之權利 (Right of company to cancel)。

32.無默示拋棄 (No implied waivers)。

33.準據法 (Governing law)。

34.模式或設計之變更 (Change of models or designs)。

35.代位清償之權利 (Subrogation rights of company)。

36.訂定契約之權利 (Authority to make agreement)。

37.經銷商之責任 (Distributor's duties)。

38.製造商或經銷商之所有權 (Title of goods to remain in manufacturer or distributor)。

39.履約保證金 (Security or bond for performance)。

40. 型錄及廣告資料之提供 (Catalogs and advertising material to be furnished)。

41.貨物品質之擔保 (Warranty of quality of goods)。

42.價格之變動 (Change of price)。

43.貨物之交付 (Delivery of goods)。

44.終止後之請求權調整 (Adjustment of claims after termination)。

45.庫存之讓售 (Consignment of stocks)。

46. 經銷商所有權之保留 (Distributor to retain ownership)。

47. 交易商對經銷商之報告 (Reports by dealer to distributor)。

48. 交易商對過時商品之返還 (Dealer's return of obsolete merchandise)。

49. 交易商非經銷商或製造商之代理人 (Dealer not to be agent of distributor or manufacturer)。

50. 製造商對經銷商員工之侵權行為不負其責任 (Manufacturer not liable for torts of distributor's employees)。

51. 契約之修正 (Amendment of agreement)。

CONTRACT

1. Obj... this Contract
1.1. ...tomer shall order and the Executor shal'
consu... der the Technical Assignment (Ap...
Contra... ...ges) of the performance of
1.2. Per... ...nment.
Technica...

2. Obligatio... ...e Parties
2.1. The Cu... ...T shall be obliged·
Contract... ...he work perform
a) to p... ...mely all nec...
b) to pro... ...o provide f...
Executor, ...e obligc...
c) if necess... ...under th...
2.2. The Executor ...e of wo...
a) to perform ...n the ful...
b) upon perfo... ...ntract ar...
c) to perform w... ...ntract an...
determined in th...

3. Procedure of Work
3.1. The Executor shall ... ...m work b... any third r...
3.2. The Executor may e... ...conditions o...
however, subject to terms ... ...pproval or notl...
...ld for in this Contract... ...pproval or notl...
...t required.... ...tion of the Report on the performed work an...
... the Report on the performed work an...
...the Customer for signing.
...and within the period in...
...days of the date of receipt of the Report an...
...ion, the performed work shall be de...
...shall be argumented and incl...
...assignment. In such

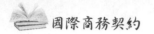

經銷商品契約

A 公司
紐約州紐約市

2000 年＿＿＿月＿＿＿日

B 公司
華盛頓州西雅圖市

　　致台端：本信函乃用以確認台端與本公司之契約如下：

第一條　本公司乃指派台端而台端乃同意擔任本公司於華盛頓州之經銷商，以供銷售物品。

第二條　台端同意盡最大之努力於前述之領域內銷售物品，且為滿足訂單之需求，台端將準備充分數量之商品於倉庫之中。

第三條　本公司同意於本契約之期間內，依當時之標準條款，銷售予台端所需求之物品，但本契約書另有規定者不在此限。

第四條　本公司對台端之物品價格，乃依接受台端訂單時，本公司所建議之芝加哥 F.O.B. 分銷價格，並減除百分之十二之折扣。發票將於運送後九十日內到期且應付款。

第五條　本契約之內容將不得解釋為賦予台端應代表本公司取得任何權利或權限，並自本公司負擔債務，或使本公司受任何拘束。

SALES DISTRIBUTION AGREEMENT

A CORPORATION

New York, N.Y.

_____ , 2000

B Company, Inc.

Seattle, Washington

Gentlemen: This will serve to confirm our agreement as follows:

Section 1. We hereby appoint you and you agree to act as our distributor in the States of Washington (hereinafter referred to as "the territory") for 〔Specify〕 (hereinafter referred to as "the goods").

Section 2. You agree to use your best efforts to promote the sale of the goods in the territory, and to maintain in your warehouse an adequate supply of the goods to enable you to fill orders for such goods in the territory.

Section 3. We agree to sell you all of your requirements for the goods during the term of this agreements, subject to our standard terms and conditions of sale then prevailing and except as otherwise specifically provided herein.

Section 4. Our prices to you will be F.O.B. Chicago based upon our suggested retail prices prevailing at the time of acceptance of your order, less a discount of 12%. Invoices shall be due and payable 90 days after shipment.

Section 5. Nothing herein contained shall be deemed to create in you any right or authority to incur any obligations on our behalf or to bind us in

第六條　非經本公司之同意，台端不得以直接或間接之方式，於前述之領域外，銷售或要約銷售本契約之物品。

第七條　本契約於二〇〇〇年＿＿＿月＿＿＿日之前，均為有效，且非經當事人於每年度契約屆滿前三個月以掛號郵件之書面通知方式，終止本契約時，本契約將繼續生效。該等終止之通知將不影響任何既存之訂購、責任或於通知前所產生之其他承諾。

第八條　本契約將優先於當事人間前已合致了解及協定，且本契約不得以口頭之方式為變更或終止，本公司以紐約州法為準據法，且依該法而為解釋之。

　　　如前述之記載為當事人間了解之正確表示，請於本信函之末端簽名。

　　　　　　　　　　　　　　　　　　　　　大　安
　　　　　　　　　　　　　　　　　　　　A 公司

　　　　　　　　　　　　　　　　　　_____
　　　　　　　　　　　　　　　　　　　　　副總裁

同意並接受：

　　　B 公司

_____
　　總　裁

any respect whatsoever.

Section 6. You shall not directly or indirectly sell or offer to sell any of the goods covered by this agreement outside the territory without first obtaining our written consent thereto.

Section 7. This agreement shall continue in force until _____, 2000 and shall continue indefinitely thereafter until terminated by either of us on any anniversary date upon not less than three months' prior written notice given to the other by registered or certified mail. Any such termination shall not affect any existing order, obligation or other commitment incurred prior to the receipt of such notice.

Section 8. This agreement supersedes all prior understandings and agreements between us, may not be changed or terminated orally, and shall be governed by and construed according to the laws of the State of New York.

If the foregoing correctly sets forth our understanding, will you please so signify by signing your name at the foot of this letter.

Very truly yours,

A CORPORATION

By _____
Vice President

ACCEPTED AND AGREED:

B COMPANY, INC.

By _____
President

§4　經銷服務契約

　　按外國廠商常要求特定廠商於一定區域內為銷售，然就該廠商之行為必須為適當之拘束，且廠商相互間之交易價格，亦必須為約定，此種契約具有買賣契約與委任契約之特質，以下就其條款說明之：

1. 銷售服務 (Distributorship Services)。
2. 期間及更新 (Term and Renewal)。
3. 經銷之地區 (Distribution Areas)。
4. 物品之價格與給付方式 (Price and Payment)。
5. 訂購之數量 (Quantity Orders)。
6. 獨家經銷權 (Exclusive Distributorship)。
7. 新產品銷售之成本 (New Product Costs)。
8. 終止 (Termination)。
9. 瑕疵物品之補正 (Correction of Defective Goods)。
10. 轉讓 (Assignment)。
11. 準據法及語言 (Governing Law and Language)。
12. 通知 (Notices)。
13. 仲裁 (Arbitration)。

CONTRACT

1. Obj... ...his Contract
1.1. ...tomer shall order and the Executor shal'...
consu... ...der the Technical Assignment (Ap...
Contra... ...ges) of the performance of...
1.2. Per... ...nment.
Technic...

2. Obligatio... ...e Parties
2.1. The Cu... ...r shall be obliged:
a) to pa... ...the work perform...
Contrac... ...mely all nece...
b) to pro... ...9 provide f...
Executor,... ...be oblige...
c) if necess... ...under th...
2.2. The Executor... ...e of wor...
a) to perform... ...the ful...
b) upon perfo... ...ntract a...
c) to perform... ...ntract...
determined in th...

3. Procedure of Work
3.1. The Executor shall ...m work t...
3.2. The Executor may en... ...any third...
however, subject to terms a... ...conditions o...
...led for in this Contract... ...pproval or noti...
...at required,... ...tion of work and within the period...
...the Report on the performed work an...
...e the Customer for signing.
...(0) days of the date of receipt of the Report and...
...ion, the performed work shall be dee...
...the Customer... ...t shall be argumented and inclu...
...sign...ment. In suc...

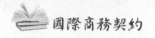

經銷服務契約

本契約於＿＿＿年＿＿＿月＿＿＿日由：
A 公司　　　　　　　　　　　　及 B 公司
地址：＿＿＿＿＿＿＿＿＿＿　　　地址：＿＿＿＿＿＿＿＿＿＿
共同簽署之。

當事人茲同意就特定商業事務為合作，而由當事人個別於特定區域為銷售，基於此合意，當事人同意條款如下：

第一條　相互經銷服務
　　　　A 公司應銷售及供應商品予 B 公司，而 B 公司應依下述契約之條款將貨品再銷售予本契約指定地點之當事人。

第二條　期間及更新
　　　　本契約有效期間為一年，於＿＿＿年＿＿＿月＿＿＿日生效之，且當事人於契約屆滿前＿＿＿日以書面通知終止契約，否則，本契約自動延長之。

第三條　當事人銷售之地區
　　　　A 公司之銷售地區為臺灣，而 B 公司銷售之地區為香港。

第四條　價格及給付
　　　　本契約商品之給付應以 F.O.B. 之價格計算之，於計算該價格時，

DISTRIBUTOR SERVICE AGREEMENT

This Agreement dated for reference the _____ day of _____ , 20 _____ .

Between: A Company

      Address: _____ (hereinafter called "A")

and   B Company

      Address: _____ (hereinafter called "B")

WHEREAS the Parties hereto have agreed in principle to cooperate in certain business matters in order that each sells and distributes in its respective country products supplied by the other. A and B, in consideration of the mutual covenants contained herein, agree as follows:

ARTICLE 1.　MUTUAL DISTRIBUTORSHIP SERVICES

A shall sell and supply to B, and B shall sell and supply to A, on the terms provided below all manner of goods as shall be agreed by them for the purpose of resale and distribution by the receiving Party within its territory of distribution as defined below.

ARTICLE 2.　TERM AND RENEWAL

The term of this Agreement shall be for one year, and shall commence on the _____ day of _____ , 20 _____ and shall be automatically renewed for a one year term thereafter on each annual anniversary date unless one or both Parties give written notice of termination to the other at least _____ (_____) days prior to the relevant anniversary date.

ARTICLE 3.　PARTIES' DISTRIBUTION AREAS

The territory of distribution of A shall be the Republic of China as governed from its capital, Taipei, and the territory of distribution of B shall be Hong Kong.

ARTICLE 4.　PRICE AND PAYMENT

The price to be paid by the receiving Party to the sending Party for all

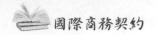

就當事人於此銷售地區之再售行為不得考量之。

第五條　無最低訂購數額

本契約之訂購數額無最低額之限制，當事人任一方得依其商業之考量而自由裁量為訂購之。

第六條　獨家銷售權

當事人任一方於特定領域有獨家之銷售權，且任一方不得於他方之領域內為直接或間接之銷售或供應。

第七條　新商品之推銷費用

A 公司應給付且負擔所有必要之成本、費用、稅捐，以介紹或推廣新商品，而 B 公司亦同時負擔與支付所有之成本、費用與稅捐，於 A 公司地域內推銷新商品。

第八條　本契約之終止

如當事人一方違反本契約之約定，任一方得於＿＿＿＿日內以書面通知終止本契約，任何一方得於下列事由發生時，隨時終止本契約：

goods ordered pursuant to this Agreement shall be paid in advance and shall be the F.O.B. price quoted by the sending Party. In calculating said price, no reference shall be made to the price to be set on resale by the receiving Party in its territory of distribution.

ARTICLE 5.　NO MINIMUM ORDERS

There shall be no requirement for a minimum quantity of goods purchased or made available for supply, nor number of orders placed under this Agreement, and A and B shall be at liberty to order from the other only such number and type of goods each requires for its business purposes in its respective territory of distribution.

ARTICLE 6.　EXCLUSIVE DISTRIBUTORSHIP

Each Party shall be the other's exclusive distributor for purposes of resale in their respective territories of distribution, and neither Party shall without the other's consent sell, supply, transfer or otherwise make available, directly or indirectly, goods over which it has control to third parties for the purpose of import or sale into the other Party's territory of distribution.

ARTICLE 7.　NEW PRODUCT INTRODUCTORY COSTS

A shall pay and be responsible for all costs, expenses and taxes incurred by its acts relating to the introduction, advertisement and transport of its new products (and samples as thereof) or goods intended primarily for advertisement purposes in B's territory of distribution. Likewise, B shall pay and be responsible for all costs, expenses and taxes incurred by its acts relating to the introduction, advertisement and transport of its new products (and samples thereof) or goods intended primarily for advertisement purposes in A's territory of distribution.

ARTICLE 8.　TERMINATION FOR CAUSE

Notwithstanding Article 14 herein, in the event of breach or default of obligation hereunder by one Party, the other may terminate this Agreement

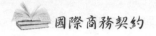

1.當事人任一方支付不夠或無法支付其到期之債務時，

2.當事人任一方基於自願或強制之解散，

3.因接獲通知為全部或一部之清算，且指定清算人者，

4.當事人任一方逾期或無法給付契約之債務者。

第九條　獨立之關係人

本契約之當事人均為獨立之契約當事人，任一方不代表、代理、受僱對方，或造成大眾認知其為合夥人。

第十條　瑕疵物品之補正

如任何一方交付予對方之貨品有瑕疵者，當事人得基於書面之請求，以運送之指示，要求對方以其費用立即為更換。

第十一條　不得轉讓

本契約之權利與義務非經當事人書面之同意，不得轉讓予第三人。

第十二條　語言

本契約以英文簽署之，如當事人就本契約有爭議時，本契約應以其一般合理之意義解釋之。

upon ＿＿＿ （＿＿＿） days written notice. Either Party may, however, terminate this Agreement immediately by written notice in the event that the other:

  i) becomes insolvent or unable to pay its debts when they fall due;

  ii) has gone into liquidation, whether voluntarily or under compulsion;

  iii) has a receiver appointed for all or a part of its undertakings or assets or has received notice of any proceeding or proposed proceeding for winding-up; or

  iv) otherwise fails or becomes unable to make payment due under this Agreement.

ARTICLE 9. INDEPENDENT CONTRACTOR

  Each Party shall carry out its business and obligations hereunder as an independent contractor and neither Party shall act or purport to act or be regarded as the principal, agent, employer or employee of the other nor shall any term herein be deemed or understood to create a partnership between A and B.

ARTICLE 10. CORRECTION OF DEFECTIVE GOODS

  In the event of defect in goods shipped by one Party hereunder, such Party, upon the written request of the other, shall immediately replace at its cost said goods with non-defective goods in accordance with shipping instructions to be forwarded by the other Party.

ARTICLE 11. NO ASSIGNMENT

  The rights and obligations arising out of this Agreement shall not be assigned or transferred to a third party without the prior written consent of the other Party to this Agreement.

ARTICLE 12. LANGUAGE

  This Agreement is executed in the English language and, in the event of a dispute between the Parties or an ambiguity arising in respect of the terms

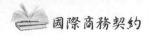

第十三條　通知

就本契約當事人之間的通知與通訊，應以親自交付、傳真等方式為交付之，除當事人以掛號郵件寄送者，該通知應為認定是營業日交付之。

第十四條　仲裁

如本契約有爭議時，當事人得終止本契約，或就本契約有終止權利之一方得請求香港或中華民國仲裁協會仲裁之。

當事人之基於授權而於＿＿＿＿年＿＿＿＿月＿＿＿＿日簽署本契約。

A 公司

＿＿＿＿＿＿＿＿＿＿＿＿＿＿
（代表人）

B 公司

＿＿＿＿＿＿＿＿＿＿＿＿＿＿
（代表人）

herein, this Agreement shall be construed in accordance with the ordinary and reasonable meanings thereof.

ARTICLE 13.　NOTICES

All notices and communications between the Parties relating to this Agreement shall be made and delivered in the English language by personal delivery, electronic facsimile by prepaid registered airmail to the Parties addresses set out above, unless notice of change of address has been duly delivered. Delivery of notices and communications hereunder shall be deemed to be made on the business day sent except in the case of prepaid registered airmail, which shall be deemed to be delivered on the third business day after date of mailing.

ARTICLE 14.　AGREED ARBITRATION

In the event of a dispute or other matter requiring resolution under this Agreement, the Parties may agree to forego their right to terminate this Agreement under Article 8 and to refer such dispute or matter to arbitration pursuant to "The Agreement between the Hong Kong Commercial Arbitration Association and the Commercial Arbitration Association of the Republic of China" executed _____ .

WHEREFORE, the Parties have caused their duly authorized signatories to execute this instrument at _____ , _____ _____ .

A   B
by: _____   by: _____

_____   _____
(Title)   (Title)

§5 租賃契約 (Lease Agreement)

　　所謂租賃契約乃出租人 (Landlord) 與承租人 (Tenant) 約定，由出租人以物租與承租人使用、收益，承租人支付租金之契約（參考民法第四二一條第一項）。租賃契約一般應約定下列事項：

1. 租金 (Rent)。
2. 租賃之標的 (Occupancy)。
3. 承租人禁止變更租賃物之義務 (Alterations)。
4. 承租人與出租人之修繕義務 (Repair)。
5. 標的物毀損之責任 (Property-loss, Damage, Reimbursement)。
6. 標的物火災之責任 (Destruction-Fire or other Cause)。
7. 權義之轉讓 (Assignment)。
8. 破產 (Bankruptcy)。
9. 債務不履行 (Default)。
10. 出租人之救濟 (Remedies of Landlord)。
11. 費用之支付 (Fees and Expenses)。
12. 租賃期間之終止 (End of Term)。
13. 通知 (Notices)。
14. 擔保 (Security)。
15. 管轄法院 (Jurisdictions)。
16. 準據法 (Governing Law)。

CONTRACT

...this **Contract**
...tomer shall order and the Executor shal'
...der the Technical Assignment (Ap
...ges) of the performance of
...nment.

1. Obj
1.1.
consu
Contra
1.2. Per
Technica

he Parties
...r shall be obliged
...he work perform

2. Obligatio
2.1. The Cu
Contract
a) to pa ...mely all nece
b) to pro ...o provide f
Executor, ...be obligee
c) if necess ...under t
2.2. The Execut
a) to perform ...e of wo
b) upon perfo ...n the ful
c) to perform ...ntract a
determined in th ...ntract a

3. Procedure of Work
3.1. The Executor shall ...rm work t
3.2. The Executor may et ...any third p
...however, subject to terms a ...conditions o
...d for in this Contract ...pproval or not
...not required. ...ion of work and within the period 1... ...work an
...e Report on the performed work an
...e the Report on the performed work a
...e the Customer for signing.
...days of the date of receipt of the Report an
...tion, the performed work shall be de
...t shall be argumented and inclu
...the assignment. In such

租賃契約

本契約乃由主營業所設於（　）之 A 為出租人及主營業所位於（　）且居住於（　），B 為承租人。

出租人同意將下列之房地出租於承租人：

（　）乃位於（　）而面積為（　），該契約之期限為一九八二年一月一日起至二〇〇一年十二月三十一日止二十年，該房地乃用於工作水電及五金行之商店，且用以銷售水電之裝備及其他供應品，本契約乃基於下列之條款：

第一條　承租人應支付予出租人至一九九五年十二月三十一日止每年九千元之房租。自一九九六年一月一日起，每年之房租乃依紐約市之消費者物價指數之增加比例而增加租金。

第二條　承租人應盡其注意管理房地，且就該建築、架構、設備及家具應以其費用為修理並保持良好狀況，承租人且不得對其為阻礙或遷離，承租人亦不得非經出租人之同意就該房地之建築與架構為變更或增加；於契約到期以後，承租人對該房地應以合理之狀態為交還；由承租人就該房地所為之建築或架構，將成為所有人之財產。

第三條　承租人須遵守聯邦、州及地方政府，或其相關部門之法規於前述期間為有關租賃標的物之變更、預防、排除任何攪擾或其他申訴事宜。

LEASE AGREEMENT

THIS AGREEMENT BETWEEN _____, having his place of business at _____ as Landlord and _____, residing at _____, and having his principle place of business at _____, _____ as Tenant.

WITNESSETH: The Landlord hereby leases to the Tenant the following premises: _____ situated on _____ and _____ for the term of twenty years to commence from the 1st day of January 1982 and to end on the 31st day of December 2001 to be used and occupied only for and as a plumbing and hardware store, and for the sale of plumbing fixtures and supplies, upon the conditions and covenants following:

Section 1. Tenant shall pay the annual rent of nine thousand ($9,000.00) dollars until the year ending December 31, 1995. From and after January 1, 1996 annual rent shall be an amount determined by calculating the percentage increase based on the Consumer Price Index for New York City.

Section 2. Tenant shall take good care of the premises and shall, at the Tenant's own cost and expense make all repairs and maintain in good condition any and all buildings, structures, equipment and appliances and will not encumber or remove same; and shall not alter or add to the building and structures on the premises without the written consent of the Landlord; and at the end or other expiration of the term, shall deliver up the demised premises in good order or condition; provided, that all buildings and structures placed on the premises by the Tenant shall become and remain property of the Landlord.

Section 3. Tenant shall promptly comply with all statutes, ordinance, rules, orders, regulations and requirements of the Federal, State, and Local

第四條　非經出租人之書面同意，承租人、其繼承人、繼受者、執行者或管理者不得轉讓本契約，或將該房地之一部為轉租或為任何之改變。

第五條　如非因承租人或承租人之代理人或受僱人之過失而致該承租標的之建築因火災或其他原因而毀損者，如該毀損乃致該承租標的或該建築之全部毀損，或出租人於合理之時間內不為重建時，該租賃契約即為終止。

第六條　承租人乃同意出租人、其代理人或其他代表人為檢驗或修理及改變承租之標的以保障該標的之安全，得於合理之時間範圍內，進入該租賃之標的。

第七條　承租人乃同意出租人或出租人之代理人為他人購買承租標的而將該標的出示予他人，承租人且同意於契約屆滿前六個月，出租人或出租人之代理人有權於承租標的之前方訂有供「出租」或「銷售」之文字，承租人不得就該標示為阻礙。

第八條　如承租之標的於契約期間遭捨棄或控制者，或租金因全部或部分遲延給付、或承租人就本契約之義務未能履行者，出租人或其代

Government and of any and all of their Departments and Bureaus applicable to said premises, for the correction, prevention, and abatement of nuisances or other grievances, in, upon, or connected with said premises during said term.

Section 4. Tenant, successors, heirs, executors or administrators shall not assign this agreement, or underlet or underlease the premises, or any part of thereof, or make any alterations on the premises, without the Landlord's consent in writting.

Section 5. In case of damages, by fire or other cause, to the building in which the leased premises are located, without the fault of the Tenant or Tenant's agents or employees, if the damage is so extensive as to amount practically to the total destruction of the leased premises or of the building, or if the Landlord shall within a reasonable time decide not to rebuild, this lease shall cease and come to an end.

Section 6. Tenant agrees that the Landlord and the Landlord's agents and other representatives shall have the right to enter into and upon premises, or any other part thereof, at all reasonable hours for the purpose of examining the same, or making such repairs or alterations therein as may be necessary for the safety and preservation thereof.

Section 7. Tenant agrees to permit the Landlord or the Landlord's agents to show the premises to persons wishing to hire or purchase the same; Tenant further agrees that on and after the sixth month, next preceding the expiration of the term hereby rented, the Landlord or the Landlord's agents shall have the right to place notices on the premises "To Let" or "For Sale", and the Tenant agrees to permit the same to remain thereon without hindrance or molestation.

Section 8. If the premises, or any part thereof shall be deserted or become vacant during the rent or any part thereof, or if any default be made

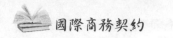

表得進入該承租之標的，出租人得進入該承租標的，承租人且放棄有關再進入之書面送達，然承租人仍應給付為承租人保留標的之相當租金。

第九條　出租人得以承租人之費用替換以承租標的之任何或全部已破裂之玻璃。就承租標的之毀損乃因承租人或承租人之代理人或受僱人之過失或不當行為所致者，承租人應以其自己之費用迅速為修補。

第十條　承租人不得就承租標的之行人道、路口或樓梯等為阻礙，並亦不得允許他人為阻礙。

第十一條　承租人不得於承租標的之全部或一部或路口設置任何標示，但經出租人書面同意者不在此限。

第十二條　出租人就因水災、電力、瓦斯、水、雨、冰、雪或任何流動或漏入該建築物所生之損害不負其責任，但其損害乃因出租人過失所致者不在此限。

in the performance of any of the convenants herein contained, the Landlord or representatives may reenter the premises by force, summary proceedings or otherwise, and remove all persons therefrom, without being liable to prosecution therefor, and the Tenant hereby expressly waives the service of any notice in writing of intention to reenter, and the Tenant shall pay at the same time as the rent becomes payable under the terms hereof a sum equivalent to the rent reserved herein.

Section 9. Landlord may replace, at the expense of Tenant, any and all broken glass in and about the demised premises. Damage and injury to the premises, caused by the carelessness, negligence or improper conduct on the part of the Tenant or the Tenant's agents or employees shall be repaired as speedily as possible by the Tenant at the Tenant's own cost and expense.

Section 10. Tenant shall neither encumber nor obstruct the sidewalk in front of, entrance to, or halls and stairs of the premises, nor allow the same to be obstructed or encumbered in any manner.

Section 11. Tenant shall neither place, or cause or allow to be placed, any sign or signs of any kind whatsoever at, in or about the entrance to the premises or any other part of same, except in or at such place or places as may be indicated by the Landlord and consented to in writing by the Landlord.

Section 12. Landlord is exempt from any and all liability for any damage or injury to person or property caused by or resulting from stream, electricity, gas, water, rain, ice or snow, or any leak or flow from or into any part of the building or from any damage or injury resulting or arising from any other cause or happening whatsoever unless the damage or injury be caused by or be due to the negligence of the

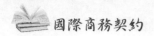

第十三條　如因承租人違反本契約時，出租人得進入該承租之標的，且得重新占有並使用該標的。承租人且拋棄有關出租人進入該標的之書面通知程序。

第十四條　本契約不得為承租標的之留置權，且不得對抗現存或將來設置之抵押權。該設置抵押權有優先於本承租契約之權利，承租人且同意為使該抵押有優先之效力應配合履行相關之程序，如承租人拒絕履行該等文件時，出租人或其指定之代表人有權終止本契約。

第十五條　承租人將於本日交付予出租人三千五百美金以供其遵守本契約條款之擔保，如承租人已履行其契約之條款或義務時，出租人應返還該擔保。如出租人依本契約為善意之出售予第三人時，出租人得將該擔保轉讓於受讓人，該出租人且對該擔保不負其責任，承租人同意應向新出租人請求返還擔保，且該標的為轉讓時，其應向另一新出租人請求，該擔保無須支付利息。

Landlord.

Section 13. If default be made in any of the covenants herein contained, then it shall be lawful for the Landlord to reenter the premises, and the same to have again, repossess and enjoy. Tenant hereby expressly waives the service of any notice in writing of intention to reenter.

Section 14. This instrument shall not be a lien against the premises in respect to any mortgages that are now on or that hereafter may be placed against the premises. The recording of such mortgage or mortgages shall have preference and precedence and be superior and prior in lien of this lease, irrespective of the date of recording, and the Tenant agrees to execute any such instrument without cost, which may be deemed necessary or desirable to further effect the subordination of this lease to any such mortgage or mortgages, and a refusal to execute such instrument shall entitle the Landlord, or the Landlord's assigns and legal representatives to the option of cancelling this lease without incurring any expense or damage and the term hereby granted is expressly limited accordingly.

Section 15. The Tenant has this day deposited with the Landlord the sum of $3,500.00 as security for the full and faithful performance by the Tenant of all the terms, covenants and conditions of this lease upon the Tenant's part to be performed, which sum shall be returned to the Tenant after the time fixed as the expiration of the term herein, provided the Tenant has fully and faithfully carried out all the terms, covenants and conditions on Tenant's part to be performed. In the event of a bona fide sale, subject to this lease, the Landlord shall have the right to transfer the security to the vendee for the benefit of the Tenant and the Landlord shall be considered released by the Tenant from all liability for the return of such security; and

第十六條　非經出租人之書面同意，承租人不得將該擔保設定抵押、轉讓或其他之阻礙。

第十七條　如該承租之標的遭廢棄或控制、承租人遲延給付租金、承租人未經出租人之同意將標的轉讓、出售或設定抵押、或對本契約之條款有所違背、或就任何聯邦、州或地方政府或其部門之法規有所違背、或承租人遭申請破產或為債權人之利益轉讓其權利時，出租人得以五日之通知要求終止本契約，依出租人所訂之日期該契約即為終止，其通知須寄往承租人現使用之標的地址。

the Tenant agrees to look to the new Landlord solely for the return of the security, and it is agreed that this shall apply to every transfer or assignment made of the security to a new Landlord. The security deposit shall bear no interest.

Section 16. The security deposited under this lease shall not be mortgaged, assigned or encumbered by the Tenant without the written consent of the Landlord.

Section 17. It is expressly understood and agreed that in case the demised premises shall be deserted or vacated, or if default be made in the payment of the rent or any part thereof as herein specified, or if, without the consent of the Landlord, the Tenant shall sell, assign, or mortgage this lease or if default be made in the performance of any of the covenants and agreements in this lease contained on the part of the Tenant to be kept and performed, or if the Tenant shall fail to comply with any of the statutes, ordinances, rules, orders, regulations and requirements of the Federal, State and Local Governments or of any and all their Departments and Bureaus, applicable to the premises, or if the Tenant shall file or there be filed against Tenant a petition in bankruptcy or arrangement, or Tenant be adjudicated a Bankrupt or make an assignment for the benefit of creditors or take advantage of any insolvency act, the Landlord may, if the Landlord so elects, at any time thereafter terminate this lease and the term hereof, on giving to the Tenant five days' notice in writing of the Landlord's intention so to do, and this lease and term hereof shall expire and come to an end on the date fixed in such notice as if the date were the date originally fixed in this lease for its expiration. Such notice may be given by mail to the Tenant addressed to the demised premises.

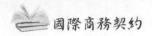

第十八條　承租人應支付予出租人依水錶或其他方法所定之有關於承租標
　　　　　的水之租金或費用。

第十九條　承租人將不會或不容許次承租人或其他人之行為，而導致其火
　　　　　災保險之費用增加。

第二十條　出租人就本契約條款嚴格執行之堅持縱有緩和時，並不表示出
　　　　　租人就其損害賠償之放棄，亦非對事後發生之違約有所放棄，
　　　　　本契約不得以口頭之方式變更、取消或終止。

第廿一條　如承租之標的因政府或準政府使用之目的而遭徵收時，本契約
　　　　　於該機構取得權利時即為終止，承租人不得就剩餘未滿之租期
　　　　　請求其償還價值，承租人亦不得就其徵收補償為請求。

第廿二條　如因承租人遲延給付租金或違反本契約之條款或於本契約期滿
　　　　　時，承租人未能將其交易設備或其他財產為最後通知前搬離時，
　　　　　前述之設備及財產視為承租人已拋棄，且成為出租人之財產。

Section 18. Tenant shall pay to Landlord the rent or charge, which may, during the demised term, be assessed or imposed for the water used or consumed in or on the premises, whether determined by meter or otherwise.

Section 19. Tenant will not nor will the Tenant permit undertenants or other persons to do anything in the premises, or bring anything into the premises, or permit anything to be brought into the premises or to be kept therein, which will in any way increase the rate of fire insurance on the demised premises.

Section 20. The failure of the Landlord to insist upon a strict performance of any of the term, conditions and covenants herein, shall not be deemed a waiver of any rights or remedies that the Landlord may have, and shall not be deemed a waiver of any subsequent breach or default in the terms, conditions and covenants herein contained. This instrument may not be changed, modified, discharged or terminated orally.

Section 21. If the whole or any part of the demised premises shall be acquired or condemned by eminent domain for any public or quasi public use or Purpose, then and in that event, the term of this lease shall cease and terminate from the date of title vesting in such proceeding and Tenant shall have no claim against Landlord for the value of any unexpired term of the lease. No part of any award shall belong to the Tenant.

Section 22. If after default in payment or rent or violation of any other provision of this lease, or upon the expiration of this lease, the Tenant moves out or is dispossessed and fails to remove any trade fixtures or other property prior to such default, removal, expiration of lease, or prior to the issuance of the final order or execution of

第廿三條　如因出租人之再進入租賃標的而終止租約、或承租人進行簡易程序、或承租人放棄該承租標的時，承租人仍應負其責任，且承租人仍應按月給付租金，承租人且同意須賠償因違反本契約所生之損害。

第廿四條　承租人同意放棄有關依（　　）之法律有關贖回所有權利。

第廿五條　如出租人乃因政府之緊急動員、政府或其相關部門之法規或物品因戰爭或緊急狀況無法供應時，縱出租人無法或遲延提供服務或為任何之修理變更相關設備或器具時，承租人仍應支付租金及遵守本契約之條款。

the warrant, then and in that event, the fixtures and property shall be deemed abandoned by the said Tenant and shall become the property of the Landlord.

Section 23. In the event that the relation of the Landlord and Tenant may cease or terminate by reason of the re-entry of the Landlord under the terms and covenants contained in this lease or by the ejectment of the Tenant by summary proceedings of otherwise, or after the abandonment of the premises by the Tenant, it is hereby agreed that the Tenant by summary proceedings or otherwise, or after the abandonment of the premises by the Tenant, it is hereby agreed that the Tenant shall remain liable and shall pay in monthly payments the rent which accrues subsequent to the re-entry by the Landlord, and the Tenant expressly agrees to pay as damages for the breach of the covenants herein contained.

Section 24. Tenant waives all rights to redeem under any law of the State of _____ .

Section 25. This lease and the obligation to Tenant to pay rent hereunder and perform all of the other covenants and agreements hereunder on part of Tenant to be performed shall in nowise be affected, impaired or excused because Landlord is unable to supply or is delayed in supplying any service expressly or impliedly to be supplied or unable to make, or is delayed in making any repairs, addition, alterations or decorations or is unable to supply or is delayed in supplying any equipment or fixtures if Landlord is prevented or delayed from so doing by reason of governmental preemption in connection with a national emergency or in connection with any rule, order or regulation of any department or subdivision thereof of any governmental agency or by reason of the

第廿六條　如因修理或改良承租標的之建築或設備，或因未遵守相關主管機關之法令而致承租人使用標的不便時，承租人不得主張降低或減少租金或其他補償。

第廿七條　出租人就前一承租人或其他的不法占有承租標的致無從按契約生效交付標的予承租人時，出租人將不負其責任，然該租金非於將標的物交付於承租人不得計算，惟租期並非延長。

第廿八條　承租人應以其費用至保險機構取得公共機構之保險單，該保險單應指定出租人為受益人且其金額不得少於五十萬元，該保險單須補償因該標的物所致之其他個人損害或財產損害之訴訟或請求。

第廿九條　承租人應支付於該標的營業所生之執照費、營業及銷售稅捐，承租人且應負擔因使用自來水、下水道、電力、電話、暖氣及其他電力所生之費用。

第三十條　除第一條之租金外，承租人應支付出租人之有關市、州或其他稅務機關不動產稅捐，出租人應於收到稅捐之通知後限數通知承租人，承租人於出租人收到租稅通知後次月第一日即應支付

condition of supply and demand which have been or are affected by war or other emergency.

Section 26. No diminution or abatement of rent, or other compensation, shall be claimed or allowed for inconvenience or discomfort arising from the making of repairs or improvements to the building or to its appliances, nor for any space taken to comply with any law, ordinance or order of a governmental authority.

Section 27. Landlord shall not be liable for failure to give possession of the premises upon commencement date by reason of the fact that premises are not ready for occupancy or because a prior Tenant or any other person is wrongfully holding over or is in wrongful possession, or for any other reason. The rent shall not commence until possession is given or is available, but the term herein shall not be extended.

Section 28. Tenant shall, at his own cost and expense, obtain a public liability policy from an insurance carrier satisfactory to the Landlord, and designating the Landlord as a named insured in an amount not less than $500,000.00, to insurer and indemnify the Landlord against all claims, suits and demands for personal injury or property damage occurring on the premises and shall furnish to the Landlord a certificate of such insurance.

Section 29. Tenant shall pay all license fees, business and sales taxes arising from the business conducted on the premises and all charges and assessments for water, sewer rent, gas, telephone, electricity, heat and power applicable to the premises during the term of this lease.

Section 30. In addition to the rent set forth in Section 1, Tenant shall pay to Landlord all real estate taxes assessed against the property by the City, County or State of ＿＿＿ or any other taxing authority. The

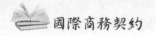

前述之費用。

第卅一條　本租賃契約將優於且取消當事人就租賃所為之其他口頭或書面合意，但依前所合意所生之責任則不在此限。

第卅二條　以下之設備（如列表）為出租人之財產，且非經出租人之書面同意不得任意移除。
　　本承租契約之承諾及合意乃當事人所同意，該合意乃拘束雙方當事人、受讓人、繼承人、執行人及管理人。

　　當事人乃於 2000 年 _____ 月 _____ 日交換簽字且以印章與（　）簽訂此契約。

Landlord shall notify the Tenant each year upon receipt of the tax bill and Tenant shall pay the Landlord the amount shown promptly and not later than the first day of the month following receipt of such notice.

Section 31. This lease supersedes and cancels any prior oral or written agreements between the Landlord and Tenant on and as of the day preceeding the commencement of the term hereof, except as to any liability or obligations which may or shall have accrued under and by virtue of any prior agreement prior to such termination date.

Section 32. The following equipment is the property of the Landlord and is not to be removed without the written consent of the Landlord: (list)

AND IT IS MUTUALLY UNDERSTOOD AND AGREED that the covenants and agreements contained in the within lease shall be binding upon the parties and upon their respective successors, heirs, executors and administrators.

IN WITNESS WHEREOF, the parties have interchangeably set their hands and seals (or caused these presents to be signed by their proper corporate officers and caused their proper corporate seal to be hereto affixed) this day of ＿＿＿,2000.

§6 獨家代理合約

國外知名商品於臺灣銷售，通常必須請有銷售或商業推廣經驗之廠商為代理廠商，此種契約所包括之約定包括當事人間之推廣、物品交易、代理行為之限制等，其具有委任、銷售或代辦商之契約性質，其主要條款如下：

1. 獨家代理權 (Exclusive Right)。
2. 獨家代理之範圍 (Territory)。
3. 契約之有效期間 (Term)。
4. 獨家代理之經營行為 (Business Operations)。
5. 利益分配 (Compensation)。
6. 契約之轉讓 (Assignment)。
7. 商號名稱 (Trade Name)。
8. 違約及終止 (Default and Termination)。
9. 救濟 (Remedies)。
10. 不可抗力 (Force Majeure)。
11. 仲裁 (Arbitration)。
12. 通知 (Notice)。
13. 保密 (Confidentiality Obligation)。
14. 準據法 (Governing Law)。

CONTRACT

1. Obj ... this Contract

1.1. ... tomer shall order and the Executor shall ... consu ... der the Technical Assignment (Ap ...
Contra ... ages) of the performance of ...
1.2. Per ... nment.
Technica ...

2. Obligatio ... he Parties

2.1. The Cu ... r shall be obliged ...
a) to pa ... the work perform ...
Contrac ... mely all nece ...
b) to pro ... o provide f ...
Executor, ... be oblige ...
c) if necess ... under th ...
2.2. The Executo ... e of wor ...
a) to perform ... the ful ...
b) upon perfor ... ntract a ...
c) to perform ... ntract a ...
determined in th ...

3. Procedure of Work

3.1. The Executor shall ... m work ...
3.2. The Executor may e ... any third ...
however, subject to terms ... conditions o ...
... led for in this Contract ... pproval or not ...
... not required. ... tion of work ... e the Report on the performed work and ...
... the Customer for signing.
... days of the date of receipt of the Report ...
... tion, the performed work shall be d ...
... assignment. In suc ...

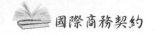

獨家代理合約

　　本契約於＿＿＿年＿＿＿月＿＿＿日，由依中華民國法律於＿＿＿設立之 A 公司，與依香港法律設立於＿＿＿之 B 公司所簽訂。

　　A 公司乃對投資人或消費者提供＿＿＿之公司，B 公司乃提供＿＿＿（「商品」）以＿＿＿為其商號名稱（「商號」）之公司。

　　基於前述之當事人合意，當事人同意下列條款：

第一條　獨家權利

　　　1. B 公司茲授權 A 公司，且 A 公司同意接受此獨家之權利，就本契約所述之範圍，提供推廣及進行運作順利之相關活動，A 公司非經 B 公司之書面同意，不得使用任何銷售代理人或其他類似之單位。

　　　2. B 公司授權 A 公司，且 A 公司接受為獨家廣告或推廣活動之商號，就有關此等廣告或推廣活動之細節，由當事人另行議定之。

　　　3. 於本契約之期限或延長之期限，A 公司不得推廣其他類似之商品，且 B 公司不得指派其他代表人於指定之地域內推廣其商品。

　　　4. A 公司同意盡其最大努力推廣該商品。

EXCLUSIVE AGENT AGREEMENT

THIS AGREEMENT is made and entered into as of the _____ day of _____ 20 _____ by and between:

A Inc., a company duly organized under the laws of the Republic of China ("ROC") having its head office located at _____ , Taipei Taiwan ("A"); and

B Inc., a company duly organized and existing under the laws of Hong Kong, having its head office located at _____ ("B").

WHEREAS A is a company organized to provide enterprise management consulting services to investors and consumers in the ROC; and

WHEREAS B is a company that among other services, _____ ("Product") using the name _____ ("Trade name");

NOW, THEREFORE, in consideration of the promises and mutual covenants contained herein, the parties hereby agree as follows:

ARTICLE 1.　EXCLUSIVE RIGHT

1. B hereby grants A, and A hereby accepts, the exclusive right to provide promotional activities and smoothing operations regarding the Product to existing and potential users of the Product in the Territory (as hereafter defined). A shall not use any sales agent or similar unit without prior written consent of B.

2. B hereby grants A, and A hereby accepts, the exclusive right to use the Tradename for purposes of advertising and promoting the Product; the details of such advertising and promoting and the function of each party with respect thereto shall be agreed between A and B.

3. During the term of this Agreement and any successive terms thereafter, A shall not promote other services similar to the Product, and B shall not appoint other representatives to promote the Product in the Territory.

4. A shall use its best efforts to promote the Product.

第二條　地域

A 公司所屬獨家之領域為臺灣。

第三條　期間

1.本契約於前述載明之日期始生效，且至該日期＿＿＿＿年內繼續有效,該契約除非經當事人於契約到期日前＿＿＿＿日至＿＿＿＿日止，為書面之通知終止此契約，否則，該契約自動延長一年。

2.依本契約第 9.4 條之規定，本契約亦為終止。

第四條　商業活動

1.A 公司就 B 公司授權為有關商品之商業活動，必須依法進行之，本契約不得排除 A 公司依中華民國之法律必須遵循之商業規範。

2.A 公司就該產品之申請應進行相關表格之提供及債信之查核，A 公司就其信用查核後，相關申請文件應送予 B 公司。B 公司乃依據 A 公司之建議而發行其商品予申請人，如 B 公司拒絕其申請時，B 公司應對 A 公司說明其理由。

3.A 公司就產品使用者之問題應盡其可能予以說明。

4.A 公司應協助 B 公司就使用人到期日之債務予以求償，如客戶遲延給付者，A 公司應協助 B 公司為必須之行動以請求債務之清償。

ARTICLE 2.　TERRITORY

　　A's exclusive area ("Territory") shall be Taiwan.

ARTICLE 3.　TERM

1. This Agreement shall come into effect on the date first above written and shall remain in full force and effect for a term of _____ (_____) years from such date. Thereafter this Agreement shall be automatically extended for further successive terms of One (1) year each unless written notice of termination is received by the party not giving notice at least _____ (_____) days but not more than _____ (_____) days prior to the end of any term.

2. Notwith standing Article 3, subclause 1, this Agreement shall terminate upon fulfillment of the conditions specified in Article 9, subclause 4.

ARTICLE 4.　BUSINESS OPERATIONS

1. A shall conduct all business operations with respect to the Product to the extent authorized by B and permitted under the laws of the ROC. Notwithstanding any other provision of this Agreement. A shall not be required to perform any business operation specified herein which is contrary to the laws and regulations of the ROC or beyond the business scope of A.

2. A shall provide forms to applicants, and conduct credit checks with respect to applications for the Product. Following credit checks by A, applications shall be transferred to B.

   B shall issue the Product to applicants based upon the recommendation letter of the general manager of A. B has a right to refuse to issue the Product. Upon such refusal, B shall disclose reasons for the refusal to A.

3. A shall handle in the Territory inquiries from users of the Product as far as possible.

4. A shall make every effort to assist B to collect money at the due date from users of the Product. If any user delays payment, A shall assist B to take such actions as are appropriate to collect the debt.

5. A 公司就使用者所須之服務，應包括飯店、租車保留、旅行、財務諮詢及登記卡片登記遺失、姓名及地址變更、及定期通知之服務等。

6. A 公司應提供願意接受該商品之商店或商人必要之聯絡服務。

7. A 公司應以 B 公司所指定之本地銀行合作，基於指定銀行及法律之許可，A 公司將對該卡片的持有人收取給付，並支付此等銀行有關信用卡使用人之給付，就接受此商品之商人或商店應予以協助或進行債務之求償。

第五條　報酬

於本契約有效期間，於前述本契約領域中因契約運作所產生的淨利或損失，應移轉予 C 公司，

(1) B 公司之費用及 A 公司之服務費，應以誠信原則為計算。

(2) B 公司之費用，應包括所有有關 C 公司信用卡業務之費用，包括：
①於香港發行信用卡之費用，
②於香港進行資料及維持之成本，
③於香港郵寄及通訊之費用，
④於香港進行工作運作資金之利息費用，
⑤於香港差旅及其他之成本，
⑥B 公司於臺灣發生之費用。
(3) 當事人同意於香港進行之各項業務，於中華民國法律許可範圍

5. A shall provide necessary services to users of the Product, including hotel and rent-a-car reservations, travel and financial consulting services, as well as lost and stolen card registration, address and name change procedures and periodical publication services.

6. A will provide liaison services to shops and merchants which wish to accept the Product in the Territory.

7. A will work with a local designated bank selected by B with A's assistance. To the extent permitted by law and approved by the designated bank, A will collect payments from card users for remittance to such bank and collect and process debit notes from shops and merchants which accept the Product in the Territory.

ARTICLE 5.　COMPENSATION

　　The net profit or loss of operations with respect to the Territory accumulated during the validity of this Agreement shall be transferred to C on the effective date of the Exclusive Rights Agreement specified in Article 6, subclause 2 hereof. A's "service fee" shall mean A's out-of-pockets expenses, promotion costs and commissions to sales agents with respect to the Product.

(1) B's out-of-pockets expenses and A's service fee shall be calculated in good faith on a costs basis.

(2) B's out-of-pockets expenses shall include all costs related to C business in the Territory as follows:

　　1) card issuing costs in Hong Kong,

　　2) data processing and maintenance costs in Hong Kong,

　　3) mail and telecommunication costs in Hong Kong,

　　4) interest expenses on working funds in Hong Kong, and

　　5) travel and other costs advanced in Hong Kong,

　　6) costs borne by B in Taiwan.

(3) It is the intention of the Parties to transfer from Hong Kong to the Territory

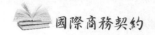

之內應移轉至臺灣。

　⑷ A 公司之服務費，應由 B 公司按月給付之。

第六條　轉讓

　1.本契約、及本契約之各項條款應拘束當事人及其受讓人，但本契約之任何權利不得直接或間接非經當事人同意轉讓與他人。

　2.當事人茲承諾且同意依中華民國法律設立 C 公司，且使 C 公司由 B 公司取得獨家權利之契約，A 公司應依本契約將其所有之權利與義務於前述契約有效期間內轉讓與 C 公司。

　3.該轉讓之權利與義務應符合下列事項：

　　⑴ 應包括但不限於因 A 公司為本契約之當事人、代理人或代表而依本契約進行之各項服務所生之請求或義務；但

　　⑵ 該權利及義務並不應包括 A 公司於權利移轉前所生之服務費。

第七條　當事人之關係

　A 公司為依本契約之條款，就其營運為獨立之當事人並具有完全

business operations which are conducted in Hong Kong at the initial stage of the business to the extent permitted under the laws of the ROC.

(4) A's service fee shall be paid monthly by B.

ARTICLE 6.　ASSIGNMENT

1. This Agreement and each and every covenant, term and condition of this Agreement shall be binding upon and inure to the benefit of the Parties and their respective successors, but neither this Agreement nor any rights under this Agreement shall be assignable directly or indirectly by either Party without the prior written consent of the other party.

2. The parties acknowledge and agree that it is their intention to establish a company, C, under the laws of the Republic of China, and to cause such company to enter into an Exclusive Rights Agreement with B. Notwithstanding Article 6, subclause 1, A shall transfer all of its rights and obligations under this present Agreement to C on the effective date of said Exclusive Rights Agreement.

3. The transfer of rights and obligations specified in subclause 2 hereof:

   a) Shall include but not be limited to the complete transfer of all the claims and obligations which B shall have with respect to services provided by A as a Party to this Agreement or as the agent or representative of B; but

   b) Notwithstanding any other provision of this Agreement, such rights and obligations shall not include A's service fee accruing prior to the transfer specified in Article 6, subclause 2 or B's accounts receivable from A for out-of-pockets expenses incurred as a result of assisting A's sales activities prior to the transfer specified in Article 6, subclause 2; such service fee and accounts receivable to remain the rights and obligations of A.

ARTICLE 7.　RELATIONSHIP OF PARTIES

A is an independent contractor with entire control and direction of its

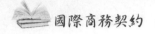

之控制權。

第八條　商號

1. B 公司對商號有完全及獨家之權利，B 公司乃授權 A 公司於本契約之有效期間內，推廣該商號。

2. A 公司將協助 B 公司於臺灣保護該商號之權利，B 公司將免除 A 公司應使用該商號而導致第三人之請求。

第九條　債務不履行及終止

1. 除本契約另有規定外，本契約得由當事人同意終止之。

2. 如當事人另一方未能履行本契約之義務時，另一造得指名其違約之事實，通知該違約之當事人。

3. 如違約之當事人之違約屬重大,且對於收到通知於＿＿＿日內未改正，該通知之當事人得告知該違約之當事人終止本契約。

4. 就 C 公司以 B 公司之獨家權利契約，依本契約之決定得自動終止之。

5. 於契約終止後，當事人就明示或意圖於本契約終止仍為有效之內容，就此部分仍屬有效，該終止並不妨礙非違約之一方之請求，亦不得認為使違約之一方於終止前之給付金錢義務已為消滅。

business and operations, subject only to the conditions and obligations of this Agreement. No agency, employment or partnership is created by this Agreement.

ARTICLE 8.　TRADE NAME

1. The Trade name is and shall remain the exclusive property of B. B hereby grants A the right to use the Trade name to promote the Product during the term of this Agreement.

2. A will assist B to protect its proprietary interests in the Trade name in the Territory. B will indemnify and hold A harmless from all claims by third parties arising from A's and B's use of the Trade name in the Territory.

ARTICLE 9.　DEFAULT AND TERMINATION

1. This Agreement, subject to the provisions contained herein, may be terminated by the mutual agreement of the parties hereto.

2. In the event either party hereto fails to perform its obligations hereunder or comply with the terms and conditions of this Agreement, the other party may issue a notice to the defaulting party specifying the breach or default and stipulating that such breach or fault be rectified.

3. In the event that the breach or default is material and is not rectified within _____ (_____) days from the receipt of notice of breach by the defaulting party, the party giving notice may terminate this Agreement forthwith by giving notice to the defaulting party.

4. This Agreement shall automatically terminate upon the effective date of the Exclusive Rights Agreement between C and B as specified in Article 6, subclause 2.

5. Following termination, the parties shall still be subject to the provisions of this Agreement as are expressed or intended to survive, operate or have effect after such termination. Termination shall not prejudice any claim of the non-defaulting party and shall not relieve the defaulting party of the obligation to pay monies owed under this Agreement arising from actions

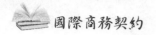

第十條　放棄及其他救濟

　　　1.對當事人之一方為請求或確實履行本契約之義務，不構成本契約履行之拋棄，而特定之拋棄，不構成事後違約之其他權利之放棄。

　　　2.當事人就本契約有關救濟權利之主張，不構成當事人就本契約之終止。

第十一條　不可履行之條款

　　　本契約任何條款之無效、非法或無法履行，當事人應以最大之努力使本契約之效果獲得實現，但如該等條款為本契約有效之基本條件者，該等替代方法必須由當事人同意採行之。

第十二條　不可抗力

　　　1.任意一方因不可抗力之事由而未能履行契約者,不構成違約，但任一方應以其最大之努力採行所有之行動，使當事人符合本契約之條款。

　　　2.除因事件之性質無法進行者，任何因不可抗力受損害之當事人應於事件後＿＿＿＿日內以書面通知對造，並盡其最大之努力排除該等損害或為救濟。

taken prior to termination.

ARTICLE 10.　WAIVER/OTHER REMEDIES

1. Failure of a party to insist upon the strict and punctual performance of any provision of this Agreement shall not constitute waiver of, or estoppel against, asserting the right to require such performance, nor shall a waiver or estoppel in one case constitute a waiver or estoppel with respect to a later breach whether of a similar nature or otherwise.

2. Nothing in this Agreement shall prevent a party from enforcing its rights by such remedies as may be available in lieu of termination.

ARTICLE 11.　UNENFORCEABLE TERMS

In the event any term or provision of this Agreement shall for any reason be invalid, illegal or unenforceable in any respect, both parties shall make best efforts to maintain the intended effect of this Agreement by adopting alternative measures. However, in the event that said term or provision is essential to the viability of this Agreement, alternative measures may be adopted only upon the agreement of both parties.

ARTICLE 12.　FORCE MAJEURE

1. The failure or delay of a party to perform any obligation under this Agreement solely by reason of acts of God beyond its control shall not be deemed to be a breach of this Agreement; provided, that the party so prevented from complying with this Agreement shall continue to take all actions within its power to comply as fully as possible with this Agreement.

2. Except where the nature of the event shall prevent it from doing so, the Party suffering such Force Majeure shall notify the other party in writing within _____ (_____) days after the occurrence of such Force Majeure and shall in every instance, to the extent it is capable of doing so, use its best efforts to remove or remedy such cause with all reasonable dispatch.

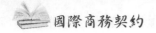

第十三條　仲裁

本契約所生之爭議、主張或契約之違約事項，應以仲裁方式解決之。

第十四條　通知

1.本契約必須所為之通知，應以書面，並加掛號或親自送達之方式送達前述當事人所記載之地址。

2.前述通知於對照收受該通知後生效。

3.當事人得依前述之規定變更其地址。

第十五條　保密之義務

1.除中華民國法律有所規定，或任何主管機關對當事人有管轄權者，或本契約有特別之規定者，當事人就本契約之資訊應予以嚴格之保密。

2.除為本契約之利益或目的，當事人任何一方不得利用該等資訊而謀利。

3.前述之義務於下列狀況不適用之：

⑴前述之資訊於契約簽訂之日前取得者，

⑵該等資訊已為公告周知者，

⑶該等資訊因出版或其他方式，而非屬當事人一造之過失成為公共資訊者，

⑷該等資訊是由第三人經合法之方法取得，且該第三人並非

ARTICLE 13.　ARBITRATION

Any controversy or claim arising out of or in relation to this Agreement, or reach of this Agreement, shall be finally settled by arbitration.

ARTICLE 14.　NOTICE

1. Any notice required or permitted to be given under this Agreement shall be made in writing and delivered by registered mail or served by personal delivery to the respective addresses of the Parties first given above.

2. The notice specified in subclause 1 shall become effective upon receipt by the respective parties.

3. Either party may change its address for purposes of this Agreement by written notice to the other party pursuant to subclauses 1 and 2.

ARTICLE 15.　CONFIDENTIALITY OBLIGATION

1. Except as may otherwise be required by the applicable laws of the ROC and any other jurisdiction having authority over either party, or as may otherwise be provided in this Agreement, each party agrees to keep in strict confidence any information conveyed to, or acquired by, them from the other party.

2. No such information shall be used for the benefit of anyone not a party or for the purposes of any Party, except for the purpose of furthering the interests and objectives of this Agreement.

3. The foregoing obligations do not, however, apply where:

   a) the information was known to the person receiving it prior to the date thereof, and was not obtained or derived under this Agreement or any other agreement contemplated herein;

   b) the information has already been, at the time of disclosure, in the public domain;

   c) the information, which after disclosure hereunder, becomes part of the public domain, by publication or otherwise, through no fault of the person receiving the said information; or

   d) the possession of such information is obtained from a third party in lawful

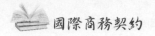

本契約應保守秘密之義務人。

第十六條　語言
本契約以英文為協商、記錄及簽署，且其應以合理、普通之意
義為解釋之。

第十七條　準據法
本契約以中華民國法律為準據法。

第十八條　契約之全部
1.本契約優先適用於簽約前之表達、說明、合議（不論其口頭
或書面）之任何協定。本契約依當事人之關係，包含當事人
所認知之整體內涵。

2.本契約所包括之條款，與訂約前之口頭或書面之證據不相抵
觸，或與訂約時同時之口頭或書面之協定不相違背。

3.非經當事人以書面且簽署之方式，本契約不得修正，且於必
要時應經主管機關之核准。

4.本契約之標題為當事人理解之必要，而不構成或影響本契約
之條款。
基於前述，當事人於前述日期簽署本契約。

A 公司　　　　　　　　　　　　B 公司

_____　　　_____
（由 _____ 簽署）　　（由 _____ 簽署）

（職位）　　　　　　　　　　　（職位）

possession of such information, who is not under a confidentiality obligation to the person from whom the information originated.

ARTICLE 16.　LANGUAGE

This Agreement has been negotiated, written and signed in the English language, and shall be governed and construed in accordance with the reasonable and ordinary meaning thereof.

ARTICLE 17.　GOVERNING LAW

This Agreement shall be interpreted in accordance with and governed by the laws of the ROC.

ARTICLE 18.　ENTIRE AGREEMENT

1. This Agreement supersedes all previous representations, understandings, or agreements, oral or written, between the parties with respect to the subject matter of this Agreement. This Agreement contains the entire understanding of the parties as to the terms and conditions of their relationship.

2. Terms included in this Agreement may not be contradicted by evidence of any prior oral or written agreement or of a contemporaneous oral or written agreement.

3. No change, alterations, or modifications to this Agreement shall be effective unless in writing and signed by authorized representatives of both parties and, if required, upon approval by competent authorities of the ROC.

4. Headings of Articles in this Agreement are for convenience only and do not substantively affect the terms of this Agreement.

IN WITNESS WHEREOF, the parties have executed this agreement on the date first above written.

A Inc.　　　　　　　　　　　　　B Inc.

By: _____　　By: _____

Title: _____　　Title: _____

§7 管理顧問契約

按公司經營就部分或全部業務得委託第三人為之，而該等顧問之給付內容與服務等，必須有明確之依據，因此，顧問管理契約對實務上頗為重要，其契約之特質主要為委任或其他服務所形成的混合契約，其主要之條款如下：

1. 顧問公司之義務 (Duties of Management Consulting Company)。
2. 員工之指派 (Secondment of Employees)。
3. 報酬 (Compensation)。
4. 費用 (Expenses)。
5. 稅捐 (Taxes)。
6. 責任 (Liability)。
7. 期間 (Term)。
8. 陳述 (Representations)。
9. 契約之修正 (Amendment)。
10. 仲裁 (Arbitration)。
11. 通知 (Notice)。
12. 準據法 (Governing Law)。

CONTRACT

1. Obj... this Contract
1.1. ...tomer shall order and the Executor shal...
consu... ...der the Technical Assignment (A...
Contra... ...ges) of the performance of...
1.2. Per... ...nment.
Technica...

2. Obligatio... ...e Parties
2.1. The Cu... ...r shall be obliged:
a) to pa... ...mely all nec...
Contrac...
b) to pro... ...provide...
Executor, ...be oblige...
c) if necess... ...under th...
2.2. The Executo... ...e of wo...
a) to perfor... ...the ful...
b) upon perfo... ...ntract a...
c) to perform... ...ntract a...
determined in th...

3. Procedure of Work
3.1. The Executor shall... ...rm work t... ...any third...
3.2. The Executor may e... ...conditions o...
however, subject to terms... ...pproval or not...
...led for in this Contract...
...ot required. ...tion of work and within the period i... ...performed work a... ...e the Report on the performed work s... ...g the Customer for signing. ...o the Customer for signing. ...d days of the date of receipt of the Report an... ...ion. the performed work shall be dec... ...shall be argumented and inclu... ...ll the assignment. In such...

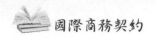

管理顧問契約

　　本契約乃由依中華民國法律所設立，主事務所於＿＿＿之 A 公司，與依＿＿＿法律所設立，主事務所為＿＿＿之 B 公司所簽訂。

源　由

　　A 公司乃進行有關研究、評估、計劃、顧問及分析等業務，而於中華民國進行該項業務。

　　B 公司乃於該項營運有卓越之聲譽及經驗。

　　茲基於前項源由，雙方當事人同意以下之條款：

第一條　B 公司之義務與責任

　　一、B 公司於本契約期間應就管理顧問之服務提供完整計劃之摘要，B 公司就其管理之策略應為說明，且就該管理之預期結果、人員配置為敘述，B 公司應就前述事項於每年年度終了時提出計劃。

　　二、為 A 公司之管理及運作改善之必要，B 公司應盡其最大努力提供以下事務之管理顧問服務：

　　⑴ A 公司之一般行政，

MANAGEMENT CONSULTING AGREEMENT

This Agreement is entered into the _____ day of _____ by and between _____, a company limited by shares existing under the laws of the Republic of China, having its head office located at _____ (hereinafter "A Company"), and _____, a company limited by shares existing under the laws of _____, having its offices located at _____ (hereinafter "B Company").

WITNESSTH

WHEREAS, A Company has been formed for the purposes of engaging in the business of research, evaluation, planning, consulting, analysis and other related business pursuant to the regulations prevailing in the Republic of China (hereinafter the "ROC"); and

WHEREAS, B Company possesses an excellent reputation and extensive experience in relation to such business.

NOW, THEREFORE, in consideration of the promises and the mutual covenants hereinafter set forth, the parties agree as follows:

ARTICLE 1. DUTIES AND OBLIGATIONS OF B COMPANY

1. B Company shall provide a brief of the entire plan regarding the management consulting services to be provided during the term of this Agreement. In addition, B Company shall also provide a statement of the management policy, an estimation of results to be achieved and a plan regarding secondment of personnel by B Company. Furthermore, B Company shall provide a plan by the end of each year for the following year with respect to the matters indicated above.

2. In order to improve the management and operation of A Company, B Company shall use its best efforts to provide management consulting and services with respect to the following items:

(1) The general administration of A Company;

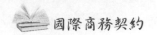

(2) A 公司年度預算之準備，

(3) 約談、聘僱及就 A 公司必要人員之僱用；決定 A 公司員工之薪資；員工之訓練；及 A 公司人員之管理服務，

(4) 建立 A 公司財務、成本控制之體系，

(5) 依一般公認會計原則維持適當之帳冊，

(6) 就與 A 公司有關係之政府及主管機關應對事宜、及對 A 公司與其他人員、事務所和公司等有相關事務之處理，

(7) 與 A 公司有關之場地與設備之管理、出租、清理、修理與維持，

(8) 為 A 公司營運之必要物品、設備、電腦、材料及供應品之買賣、出租、取得、出售、讓與、處分等其他處理等，

(9) 與 A 公司管理有關之行銷、財務、會計及行政，

(10) 有關研究、評估、計劃及分析之事務，

(11) 為 A 公司營運之必要，有關律師、會計師、技術指導員、顧問、代理人、次代理人及其他人員所提供服務之指派。

三、B 公司應善盡其義務責任，其本身或關係人應就其人員、時間、服務及相關資源隨時依管理之必要提供之。

第二條　B 公司人員之指派
一、基於本契約之規定，及為履行其義務，若需 B 公司指派其人

(2) Preparation of the annual budget of A Company;

(3) Interviewing, recruiting, and hiring all necessary personnel for A Company; determining and effecting compensation for all staff members of A Company; training the personnel of A Company; and providing personnel management services to A Company;

(4) Setting up and operating financial, treasury, and cost control systems;

(5) Maintaining proper books, vouchers, and accounting records in accordance with generally accepted accounting principles;

(6) All matters and dealings of A Company with all government, municipal, utility and other authorities and with all persons, firms, corporations and others having dealings with A Company;

(7) The management, letting, cleaning, repair and maintenance of all premises used for or in connection with A Company;

(8) Purchasing, leasing, acquiring, selling, letting, disposing, or otherwise dealing with all goods, equipment, computers, materials and supplies necessary or requisite for the business and operations of A Company;

(9) Matters relating to marketing, finance, accounting, administration and other aspects of the management of A Company;

(10) Matters relating to research, evaluation, planning and analysis;

(11) Appointment by A Company of lawyers, accountants, technical advisors, consultants, agents, sub-agents, contractors, suppliers and other persons whose services are necessary or required for the carrying on of the business of A Company.

3. B Company shall make available such of the personnel, time, services, and resources possessed by it and its affiliates as may be required to support the performance of the duties and obligations of B Company herein.

ARTICLE 2.　SECONDMENT OF EMPLOYEES FROM B COMPANY

1. If the provision of the management consulting services pursuant to this

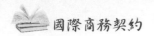

員於中華民國提供相關之服務，B 公司應提供必要之服務，
A 公司應提供協助以確保 B 公司及其所指派人員得依中華
民國法律履行其義務，該等協助包括：

  1.指派人員於相關主管機關之登記，
  2.任何有關指派人員取得工作許可、入境許可、專業人員之
    簽證等必要之協助，

二、就相關人員之指派，其條件應經當事人同意之。

第三條　報酬

一、A 公司就 B 公司（或其關係企業）所提供之服務分成兩部分
   付費：

  1.每年預付基本費新臺幣＿＿＿元，
  2.依每年管理之利潤之百分之＿＿＿收取績效費，而該績效
    費之收取不論前一年是否有所虧損。

前述基本費及績效費於不足一年時依比例為計算。

二、前述基本費應於每年之＿＿＿到期給付之，而績效費應於

Agreement requires employees of B Company or of its affiliates to perform work in the ROC, then B Company may second or send such employees to A Company for the period during which they are required to render management consulting services to A Company in the ROC. A Company shall provide all necessary assistance to B Company and such employees to ensure that B Company and such employees comply with all relevant laws of the ROC relating to the employment of foreign nationals, which assistance shall include, if required:

(1) Registration of such employees with all appropriate authorities; and

(2) All actions necessary for such employees to secure any employment passes, entry visas, professional visas, or work permits which may be required.

2. The terms and conditions (including the amount and manner of payment) under which B Company is prepared to second or send any of its employees to A Company in the ROC shall be mutually agreed by both parties.

ARTICLE 3.　COMPENSATION

1. In consideration of the management consulting services provided by B Company (or its affiliates), A Company shall pay to B Company a management consulting fee consisting of two parts:

(1) A base fee of NT$_____ per year, payable in advance; and

(2) A performance-based fee of _____ percent (_____%) of pre-tax profit, payable at the end of the year. Such remuneration shall be calculated in respect of the profits of each financial year regardless of any losses in any previous year.

Both the base fee and the performance-based fee shall be calculated on a pro-rata basis for any partial year for which management consulting services are provided.

2. The above management consulting fee shall be payable in two payments,

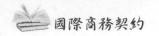
　　　　　_____前給付之。

　　三、A 公司應提供 B 公司相關之財務資訊，以解決前述給付稅捐
　　　　之義務。

　　四、A 公司給付 B 公司前述管理費以 B 公司指定之貨幣及特定
　　　　銀行為給付。

第四條　費用

　　一、除本契約另有規定者外，A 公司應以合理之原則就差旅、法
　　　　律、會計，及電腦費用，向 B 公司為給付。

　　二、⑴如 B 公司為協助 A 公司而派遣科技專家至臺灣時，A 公
　　　　　司應負擔該等專家之必要旅行與住宿費用。

　　　　⑵如該專家停留於臺灣少於_____日時，該等費用應由 B 公
　　　　　司負擔，如其於臺灣之期間超過_____日時，A 公司應給
　　　　　付 B 公司就該專家協助 A 公司期間之合理費用。

第五條　稅捐及義務

　　　　A 公司給付予 B 公司之資金，應依中華民國之法律扣繳與稅捐予
　　　　主管機關，該等稅捐應由 B 公司負擔之。

with the base fee payable in advance and due by _____ of the year respect of which the fee is being paid, while the performance-based fee shall be payable _____ .

3. A Company shall use its best endeavors to provide information to B Company to resolve any obligations to the tax authorities resulting from assessments made for income tax on profits resulting from payments made pursuant to sub-clause 1.

4. A Company shall pay the above-specified management consulting fees of B Company to such bank and in such currency as B Company requests.

ARTICLE 4.　EXPENSES

1. Except as otherwise herein provided, B Company shall be reimbursed by A Company for all fair and reasonable expenditures approved by A Company and made on behalf of A Company, including but not limited to travel, legal, accounting, and computer expenses.

2. (1) If B Company sends, or arranges to have sent, a technical expert (or experts) to Taiwan specifically to assist A Company, A Company shall bear all reasonable costs of travel and accommodation for such experts.

(2) If such expert(s) stay(s) in Taiwan to assist A Company for less than _____ days, the remuneration of such expert(s) shall be borne entirely by B Company. If such experts stay in Taiwan to assist A Company for _____ days or more, A Company shall reimburse B Company for the cost of each such expert's reasonable remuneration and benefits for the period during which he was assisting A Company.

ARTICLE 5.　TAXES AND DUTIES

All payments to be made by A Company to B Company hereunder shall be subject to deductions and withholdings for and on account of any present or future income and other taxes, levies, or assessments of any nature now and hereafter imposed, levied, collected, withheld, or assessed

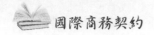

第六條　責任及補償
一、B 公司（包括子公司及其關係企業）、股東、職員、董事、受僱人或代理人不應就非故意、重大過失之行為，且非屬其應履行義務範圍內之損害受請求，該等應免責之內容包括損害賠償、費用、損失及合理之律師費等。

二、B 公司、子公司及其關係企業，應賠償及免除 A 公司及其股東因 B 公司之故意或重大過失於履行本契約之義務時，所產生造成 A 公司之缺失、損害賠償、費用、損失及合理之律師費等。

第七條　不包括之服務
一、B 公司所提供之服務不應認定為獨家，且不應認本契約所提供之服務，B 公司、子公司或關係企業不得再提供於其他公司或客戶。

二、於本契約有效期間內，其他之服務不得影響 B 公司履行本契約之義務。

第八條　期間
一、本契約於經濟部投審會就技術合作核准之日期，有效期間為

by any tax collection authority of the ROC. All such taxes, levies, or assessments shall be borne by B Company.

ARTICLE 6.　LIABILITY AND INDEMNIFICATION

1. Neither B Company (including its subsidiaries and affiliates), nor its shareholders, officers, directors, employees, or agents shall be subject to, and A Company shall indemnify and hold such persons harmless from and against, any liability for and any damages, expenses, or losses, including reasonable attorneys' fees, incurred in connection with any act or omission in the course of, connected with, or arising out of any services to be rendered hereunder, except by reason of willful misfeasance or gross negligence in the performance of the duties of B Company, or by reason of reckless disregard of the obligations and duties under this Agreement of B Company.

2. B Company, its subsidiaries and affiliates, shall indemnify and hold A Company and its shareholders from and against any liability, damages, expenses, or losses, including reasonable attorneys' fees, arising out of willful misfeasance or gross negligence in the performance of the duties of B Company, or by reason of reckless disregard of the obligations and duties under this Agreement of B Company.

ARTICLE 7.　SERVICES NOT EXCLUSIVE

1. The services of B Company hereunder are not to be deemed to be exclusive, and nothing in this Agreement shall prevent B Company, or any subsidiaries or affiliates thereof, from providing similar services to other companies and other clients.

2. Such other services and activities may not, during the term of this Agreement, interfere in a material manner with B Company's ability to meet all its obligations hereunder.

ARTICLE 8.　TERM

1. This Agreement shall become effective from the date of the approval of

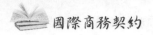

＿＿＿年，於該期間屆滿，非經當事人於屆滿前＿＿＿日為書面通知終止，該契約將自動延長＿＿＿年，於第二期間屆滿後，本契約自動延長乙期或當事人同意之期間。

二、當事人於終止前＿＿＿月為書面通知終止。

第九條　擔保

一、A 公司向 B 公司擔保，該 A 公司得依本契約與 B 公司簽約並履行其義務及行使其權利。

二、B 公司擔保下列事項：

　　⑴B 公司有其權限依本契約之條款履行義務。

　　⑵B 公司（包括其子公司及人員）有其足夠之專業知識經驗。

第十條　修訂、拘束力及轉讓

本契約得經雙方以書面同意修正之，本契約對當事人、受讓人有其拘束力，本契約非經當事人書面同意，不得轉讓之。

technical cooperation between the parties by the Investment Commission of the Ministry of Economic Affairs of the ROC and shall continue for an initial term of _____ years. After the expiration of the initial term of this Agreement, it shall automatically be renewed for an additional term of _____ years unless either party gives the other at least _____ days written notice of termination prior to the expiration of the initial term. Following expiration of the second term, this Agreement may be renewed for an additional term or terms by mutual agreement.

2. This Agreement may be terminated at any time during the initial or any subsequent term by B Company or A Company after giving the other party at least _____ months written notice of termination.

ARTICLE 9.　REPRESENTATIONS AND WARRANTIES

1. A Company hereby represents and warrants to B Company that the Corporation has full authority and power to engage B Company to handle management activities under the terms and conditions of this Agreement.

2. B Company hereby represents and warrants to A Company as follows:

(1) That B Company has full authority and power to perform its duties and exercise its rights, authorities and powers under this Agreement in accordance with the terms hereof; and

(2) That B Company, including its subsidiaries and associates, possesses sufficient professional knowledge in the relevant fields.

ARTICLE 10.　AMENDMENT; BINDING EFFECT; ASSIGNMENT

This Agreement contains the entire agreement between the Parties and may only be amended by mutual written consent. This Agreement shall be binding upon the Parties and their respective successors and assignees; provided, however that this Agreement shall not be assignable or transferable by either Party hereto without the express written consent of the other Party and any such purported assignment without written consent shall be void.

第十一條　仲裁

一、因本契約或違反本契約之任何爭議，當事人應經合議解決之，如無法經協議解決者，當事人得於＿＿＿＿日之通知請求仲裁。

二、因本契約之爭議應由仲裁解決之。

三、就本契約之仲裁應於中華民國仲裁法之規定指定三名仲裁人。

四、當事人同意就仲裁之決定及法院承認之判決為拘束。

五、當事人就法院確定中之判決之仲裁，得請求假處分。

第十二條　通知

依本契約之通知應以書面及掛號或親自送達於下列地址。

A 公司：　＿＿＿＿＿＿＿＿＿＿＿＿＿＿＿＿＿＿＿＿＿＿＿

B 公司：　＿＿＿＿＿＿＿＿＿＿＿＿＿＿＿＿＿＿＿＿＿＿＿

第十三條　解釋

本契約以英文為主，如中文與英文之內容解釋歧義時，以英文

ARTICLE 11. ARBITRATION

1. It is agreed that in case any controversy or claim arises out of or in relation to this Agreement or with respect to breach of this Agreement, the Parties shall seek to resolve the matter amicably through discussions between the Parties. Only if the Parties fail to resolve such controversy, claim or breach within _____ days of written notification by amicable arrangement and compromise may the aggrieved party seek arbitration as set forth below.

2. Any controversy or claim arising out of or in relation to this Agreement, or breach of this Agreement, shall be finally settled by arbitration.

3. The arbitration shall be conducted before three arbitrators in accordance with the Arbitration Act of the Republic of China.

4. Both Parties shall be bound by the award rendered by the arbitrators and judgment thereon may be entered in any court of competent jurisdiction.

5. Notwithstanding any other provision of this Agreement, either Party shall be entitled to seek preliminary injunctive relief from any court of competent jurisdiction pending the final decision or award of the arbitrators.

ARTICLE 12. NOTICE

Any notice to be given as required under this Agreement shall be made in writing and delivered by registered mail or served by personal delivery to the following addresses:

A Company: _____

_____

B Company: _____

_____

ARTICLE 13. INTERPRETATION

This Agreement has been written in the English language. In case of any

為準。

第十四條　準據法

本契約以中華民國法律為準據法。

基於以上合意，當事人於前述日期經合法之授權簽署本契約。

A 公司

_____

_____

（董事長）

B 公司

_____

_____

（董事長）

different interpretations between a Chinese version and the English version, the English version shall govern.

ARTICLE 14.　GOVERNING LAW

This Agreement shall be construed in accordance with and governed by the laws of the Republic of China.

IN WITNESS WHEREOF, the Parties have caused this Agreement to be executed by their respective duly authorized signatories as of the day and year first written above.

A COMPANY

By: _____

_____

Chairman

B COMPANY

By: _____

_____

Chairman

§8 開發不可撤銷信用狀申請書 (Application for Irrevocable Documentary Credit)

所謂之信用狀，不論其為「跟單信用狀」(Documentary Credit) 或「擔保信用狀」(Standby Letter of Credit)，均指開狀銀行 (The Issuing Bank) 為其本身或其客戶（即申請人，The Applicant）之請求，並依其指示所為之任何安排，於符合信用狀條款之情形下，而⑴對第三人（即受益人，The Beneficiary）或其指定人為付款，或對受益人所簽發之匯票為承兌 (Accept) 並予付款，或⑵授權另一銀行為上項付款，或對上項匯票 (Draft) 為承兌並予付款，或⑶授權另一銀行為讓購 (Negotiate)（參考信用狀統一慣例 (UCP) 第二條）。於國際貿易過程中，當買賣契約成立後，買方即向開狀銀行申請開發信用狀，以使出口商獲得信用保障。開發信用狀申請書通常包括下列事項：

1. 申請人名稱 (Name of the Applicant)。
2. 統一慣例之適用 (Application of UCP)。
3. 開狀銀行之義務 (Liability of the Issuing Bank)。
4. 審查單據之標準 (Standard for Examination of Documents)。
5. 單據有效性之免責 (Disclaimer on Effectiveness of Documents)。
6. 訊息傳送之免責 (Disclaimer on Transmission of Messages)。
7. 擔保 (Security)。
8. 連帶責任 (Joint and Several Liabilities)。
9. 條款修正或延期 (Modification of terms or Renewal of the L/C)。

CONTRACT

1. Obj... his Contract
1.1. ...tomer shall order and the Executor shall...
consu... der the Technical Assignment (A...
Contra... ...ges) of the performance of...
1.2. Per... ...ment.
Technic... ...nment.

2. Obligatio... ...e Parties
2.1. The Cu... ...r shall be obliged...
Contract... ...he work perform...
a) to pa... ...mely all nece...
b) to pro... ...o provide f...
Executor; ...be oblige...
c) if necess... ...under th...
2.2. The Executo... ...e of wor...
a) to perform... ...n the ful...
b) upon perfo... ...tract a...
c) to perform... ...ntract...
determined in... ...m work t... ...any third...

3. Procedure of Work
3.1. The Executor shall... ...conditions o...
3.2. The Executor may et... ...pproval or noti...
however, subject to terms... ...work and within the period 1...
...led for in this Contract... ...the Report on the performed work an...
...et required. ... action of work and... ...the Customer for signing...
...g the Report on the date of receipt of the Report an...
...days of the performed work shall be d...
...tion, the performed work shall be argumented and incl...
...shall... ...e assignment. In su...

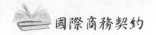

開發信用狀申請書

（申請表格：略）

上項開發信用狀之申請倘蒙　貴行核准，敝處自願遵守下列各條件：

一、本申請書確與有關當局所發給之輸入許可證內所載各項條件及細則絕對相符，並已逐一遵守，倘因申請人對於以上任何各點之疏忽致信用狀未能如期開發，　貴行概不負責。又　貴行有刪改本申請書內之任何部分，俾與輸入許可證所載者相符之權，此外申請人應遵守國際商會于一九九三年修訂信用狀統一慣例（發行物編號 500）之規定。

二、關於本信用狀下之匯票及其附屬單據等，如經　貴行或　貴行之代理行認為在表面上尚屬無訛，敝處於匯票提示時應即承兌並依期照付。

三、上項匯票單據等縱或在事後證實其為非真實或屬偽造或有其他瑕疵，概與　貴行及　貴行之代理行無涉，其匯票仍應由敝處照付。

四、本信用狀之傳遞錯誤、或遲延、或其解釋上之錯誤、及關於上述單據所載貨物、或貨物之品質或數量或價值等之有全部或一部分滅失或傳遞或因未經抵達交貨地，以及貨物無論因在海面或陸上運輸中或運抵後或未經保險或保額不足或因承辦商或任何第三者之阻滯或扣留及其他因素等各種情事，以致喪失或損害時，均與　貴行或　貴行之代理行無涉，且在以上任何情形之下該匯票仍應由敝處兌付。

APPLICATION FOR IRREVOCABLE DOCUMENTARY CREDIT

(Application Form: Omitted)

IN CONSIDERATION of your granting above request

1. This application must be in strict accordance with the conditions, specifications etc., as set forth in the import permit issued by the competent authority of the Chinese Government in this connection. The Bank shall not be held responsible for any delay in issuance of L/C due to the negligence on the part of the applicant in conformity with this request. The Bank reserves the right to alter or even delete any part or parts of this application so as to be consistent with the permit. The applicant is also requested to observe the UNIFORM CUSTOMS AND PRACTICE FOR DOCUMENTARY CREDITS 1993 revision fixed by THE INTERNATIONAL CHAMBER OF COMMERCE (Publication No. 500).

2. I/We hereby bind myself/ourselves duly to accept upon presentation and pay at your offices at maturity the drafts drawn under this Letter of Credit, if the drafts and/or accompanying documents appear in the discretion of yourselves of your agents to be correct on their face.

3. I/We agree to duly accept and pay such drafts, even if such drafts and/or documents should in fact prove to be incorrect, forged or otherwise defective, in which case no responsibility shall rest with you, and your agents.

4. I/We further agree that you or your agents are not responsible for any errors or delays in transmission or interpretation of said Letter of Credit or for the loss or non-arrival of part or of all the aforesaid documents, or the quality, quantity or value or the merchandise represented by same, or for any loss or damage which may happen to said merchandise, whether during its transit by sea or land or after its arrival or by reason of the non-insurance or insufficient insurance thereof or by whatever cause or for the stoppage, or

五、與上述匯票及匯票有關之各項應付款項，以及敝處對　貴行不論其現已發生、或日後發生經已到期或尚未到期之其他債務，在未清償以前，貴行得就本信用狀項下所購運之貨物、單據及賣得價金視同為自己所有，並應連同敝處所有其他財產：包括存在　貴行及分支機構、或貴行所管轄範圍內之保證金、存款餘額等，均任憑　貴行移作上述匯票之共同擔保，以備清償票款之用。

六、如上述匯票到期而敝處不能照兌時、或　貴行因保障本身權益認為必要時，　貴行得不經通知有權決定將上述財產（包括貨物在內）以公開或其他方式自由變賣，就賣得價金扣除費用後抵償　貴行借墊各款，毋須另行通知敝處。

七、本信用狀如經展期或重開及修改任何條件，敝處對於以上各款絕對遵守，不因展期重開或條件之修改而發生任何異議。

八、本申請書之簽署人如為二人或二人以上時，對於本申請書所列各項條款，自當共同連帶及個別負其全部責任，並負責向　貴行辦理一切結匯手續。

申請人：＿＿＿＿＿＿＿＿＿＿

　　　　（請加蓋原留印鑑）

　　　　地址：

　　　　電話：

detention thereof by the shipper or any party whomsoever, engaging myself/ourselves duly to accept and pay such drafts in all like instances.

5. I/We further agree that the title to all property which shall be purchased and/or shipped under this Letter of Credit the documents relating thereto and the whole of the proceeds thereof, shall be and remain in you until the payment of the drafts or of all sums that may be due on said drafts or otherwise and until the payment of any all other indebtedness and liability, now existing or now or hereafter created or incurred by me/us to you due or not due, it being understood that the said documents and the merchandise represented thereby and all my/our other property, including securities and deposit balances which may now or hereafter be in your or your branches' possession or otherwise subject to your control shall be deemed to be collateral security for the payment of the said drafts.

6. I/We hereby authorize you to dispose of the aforementioned property by public or private sale at your discretion without notice to me/us whenever I/We shall fail to accept or pay the said drafts on due dates or whenever in your discretion, it is deemed necessary for the protection of yourselves and after deducting all your expenses to reimburse yourselves out of the proceeds.

7. In case of extension or renewal of this Letter of Credit or modification of any kind in its terms, I/We agree to be bound for the full term of such extension or renewal, and notwithstanding any such modification.

8. In case this request is signed by two or more, all promises or agreements made hereunder shall be joint and several. I/We herewith bind myself/ourselves to settle exchange on drafts drawn under this Letter of Credit with your goodselves.

Applicant: _____

　　　　　(Authorized Signature)

　　　　　Address:

　　　　　Tel. No.:

§9 出口押匯總質權書 (General Letter of Hypothecation)

實務上所稱之「押匯」係指出口商將貨物交運後，持信用狀、提單及有關單據，前往當地國內銀行請求付款之行為，而付款之方式乃藉由匯票之讓購（Negotiate，參考 UCP 第十條）。押匯之法律性質屬授信行為，為使押匯銀行獲得擔保，乃簽訂所謂之「出口押匯總質權書」，其通常包括下列款項：

1. 申請人名稱 (Name of the Applicant)。
2. 簽訂之目的 (Purpose of the Letter)。
3. 費用之負擔 (Duty of Paying Costs)。
4. 處分之權利 (Right of Disposal)。
5. 交付單據之權利 (Right of Delivering Documents)。
6. 交付貨物之權利 (Right of Delivering Goods)。
7. 停止支付之權利 (Right of Stopping Payments)。
8. 延緩支付 (Suspend Payments)。
9. 補償條款 (Compensation Clause)。
10. 管轄 (Jurisdiction)。
11. UCP 之適用 (Application of UCP)。

CONTRACT

1. Obj... this Contract
   1.1. ...tomer shall order and the Executor shall...
   consu... der the Technical Assignment (Ap...
   Contra... ...ges) of the performance of...
   1.2. Per... ...nment.
   Technic...

2. Obligatio... ...e Parties
   2.1. The Cu... ...shall be obliged-
   a) to pe... ...he work perforr...
   Contract... ...mely all nec...
   b) to pro... ...o provide f...
   Executor; ...be oblige...
   c) if necess... ...under th...
   2.2. The Executor... ...e of wor...
   a) to perform... ...n the ful...
   b) upon perfor... ...ntract a...
   c) to perform v... ...ntract ar...
   determined in t...

3. Procedure of Work
   3.1. The Executor shall... ...rm work t...
   however, subject to terms a... ...any third...
   ...ed for in this Contract... ...conditions or...
   ...t required. ...pproval or not...
   ...tion of work and within the period 1...
   ...e the Report on the performed work an...
   ...the Customer for signing.
   ...days of the date of receipt of the Report an...
   ...ion, the performed work shall be de...
   ...sion, the performed work shall be...
   ...ssignment. In su...

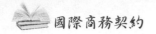

出口押匯總質權書

××銀行　台照

一、茲因立押匯總質權書人（下稱我方）為使　貴行隨時可能讓購或貼現我方所發出之押匯匯票（國內或國外）、或經我方背書之押匯匯票，爰經雙方協議：凡本書中所載各條款，均應認為永久繼續有效，隨時適用；凡我方所發出或背書之押匯匯票，無論其為直接或經手他人售與　貴行、或向　貴行貼現，均一律視同為每次讓購或貼現，並與重新簽訂本書，有同一效力。

二、茲授權　貴行或　貴行之任何經理、或代理人、或上述匯票持有人，得將（但非必需之行為）該匯票擔保品投保所有水險，並包括搶劫擄掠及岸上火災等險，所有保險費及有關費用均歸我方負擔。　貴行對擔保品享有優先受償權並得逕行處分擔保品以抵償　貴行之債權及其他有關費用或其他第三人所代付之保險等有關費用，並且不影響　貴行對其他票據債務人之請求權，同時　貴行得照普通商家代理人之事例，代我方辦理一切應辦事件，收取手續費，倘　貴行對於該指定之碼頭或倉庫並無反對之表示，我方當依照付款人或承兌人之指示，將貨物移放於公家或私人碼頭或倉庫。

三、茲授權　貴行或　貴行之任何經理或代理人或上述匯票持有人，均可

GENERAL LETTER OF HYPOTHECATION

To: ×× BANK

1. As you may from time to time negotiate or discount for me/us Bill(s) of Exchange (inland or foreign) drawn or endorsed by me/us with collateral securities, it is hereby agreed between us that the stipulations contained in this General Letter of Hypothecation shall be deemed to be continuing and effectual, shall apply to all cases in which such Bill(s) of Exchange may at any time, either directly or through other persons, be negotiated with or sold to you by me/us as if this General Letter of Hypothecation were signed by me/us on each occasion of such negotiation or discounting.

2. We authorize you, or any of your managers, or agents, or the holders for the time being of any such Bill or Bills as aforesaid (but not so as to make it imperative) to insure any goods forming the collateral security for any such Bill or Bills of Exchange against sea risk, including loss by capture, and also against loss by fire on shore, and to add the premiums and expense of such insurances to the amount chargeable to us in respect of such Bill or Bills, and to take recourse upon such goods in priority to any other claims thereon, or against us, without prejudice to any claim against any endorser or endorsers of the said Bills, for the purpose of reimbursing yourselves, or other person or persons paying the same, the amount of such premiums and expenses, and generally to take such measures and make such charges for commission and to be accountable in such manner, but not further or otherwise than as in ordinary cases between a merchant and his correspondent. And we consent to the goods being warehoused at any public or private wharf or warehouse selected by the drawees or acceptors of the Bills, unless you offer any objection to such wharf or warehouse.

3. We hereby also authorize you, or any of your managers, or agents, or the

179

接受付款人所有條件之承兌；於票據到期日票款付清後， 貴行得將隨同匯票作為擔保之附帶單據，交與付款人或承兌人。此種授權亦可適用於參加承兌，惟付款人於付款或承兌前已停止付款、或承兌前已停止支付、或宣告破產、或清理時，則應按照以下所載各款辦理。

四、茲授權 貴行：凡經 貴行或匯票承兌人代表人認為適當，在匯票到期以前無論何時， 貴行可將貨物分批交付與任何人（但非必需之行為），惟交付貨物之全部或一部分時，須收取相當金額，其金額應與發票上所開列之貨價、或與所擔保之票據所載金額成合理之比例；上述相當金額之解釋，由 貴行認定之。

五、押匯匯票經 貴行讓購後，倘因匯票或附屬單據與信用狀所規定條件不符、或其他理由而遭 貴行之貼現行或通匯行拒絕處理、或受開狀銀行拒付、或貨物在交付或其他場合被發覺貨物之品質、數量等有差異等情事、或其他任何理由致遭對方拒收時，我方願意負全責；一經 貴行通知，隨時償付 貴行匯票金額、利息與其他一切附隨費用。我方並授權 貴行：倘 貴行或 貴行之通匯銀行認為必要時，得不經通知我方， 貴行可向信用狀開狀銀行、或承兌銀行提出保證書，對此項保證，我方願意負一切責任。

holders for the time being of any Bill(s) of Exchange as aforesaid, to take conditional acceptances to all or any of such Bill, to the effect that on payment thereof at maturity, the documents handed to you as collateral security for the due payment of any such Bill(s) shall be delivered to the drawees or acceptors thereof, and such authorization shall be taken to extend to cases of acceptance for honour, subject nevertheless to the power next hereinafter given, in case the drawee shall suspend payment become bankrupt, or go into liquidation during the currency of any such Bill(s).

4. We further authorize you (but not so as to make it imperative) at any time or times before the maturity of any Bills of Exchange as aforesaid, to grant a partial delivery or partial deliveries of such goods, in such manners as you or the acceptors of such Bill or Bills of Exchange or their representatives may think desirable to any person or persons on payment of a proportionate amount of the invoice cost of such goods, or of the Bill or Bills of Exchange drawn against same. The meaning of the above-mentioned "proportionate amount" will be defined by you.

5. Should the Bill or Bills negotiated by your bank be refused handling or processing by your discounting bank or correspondent, or unpaid by issuing bank owing to some discrepancy in the Bill or Bills or the documents attached thereto with the terms and conditions of the letter of credit or for any other reasons, or should the acceptance of the shipped goods be refused because of divergence of quality, quantity, etc. of the said goods, or for any other reasons, discovered by the interested party or parties upon delivery or any other occasions, we shall take full responsibility thereof and reimburse you at any time the amount of such Bill or Bills, interest and other incidental charges incurred. We further authorize your bank to tender a letter of guarantee to the issuing bank or the accepting, etc. of the said goods, or for any other reasons, discovered by the interested party or parties upon delivery

六、茲再授權　貴行或　貴行之任何經理、或代理人、或匯票持有人，於
　　匯票提示而被承兌人拒絕承兌、或於匯票到期而被付款人拒絕支付。
　　或在票據到期前，付款人或承兌人停止支付、或宣告破產、或採取清
　　理步驟時，不論匯票是否已經承兌人附有條件承兌或絕對承兌，　貴
　　行均得將該匯票擔保品之全部、或一部分，按照　貴行或票據持有人
　　認為適當之方法，將其變賣，並將所得價款，除去通常手續費用及佣
　　金外，以之支付該票款及其匯費，倘有餘額，得由　貴行或票據持有
　　人，以之清償我方之其他票據（不論其有無擔保）、或對　貴行之欠
　　款、或對　貴行負有結算責任之其他方面欠款。凡遇保險貨物發生滅
　　失，我方授權　貴行得依照保險單取償，並扣除手續費用，與處分變
　　賣其他貨物情形同，對其所餘淨額按照上開辦法加以處理。

七、如遇匯票付款人於該匯票到期日請求　貴行或　貴行之代理銀行延緩
　　付款，而　貴行或　貴行之代理銀行認為此項請求為合理時，「得不經
　　通知我方，逕予同意延緩，我方絕無異議。」

or any other occasions, we shall take full responsibility thereof and reimburse you at any time the amount of such Bill or Bills, interest and other incidental charges incurred. We further authorize your bank to tender a letter of guarantee to the issuing bank or the accepting bank under the Letter of Credit, without any notification to us, in case your bank or your correspondent deems it fit to do so, and we solely shall be held liable for the guarantee thus offered.

6. We further authorize you, or any of your Managers, or Agents, or the Holders for the time being of any Bill(s) of Exchange as aforesaid, on default being made in acceptance on presentation or in payment at maturity, of any of such Bill(s) or in case of the Drawees or Acceptors suspending payment, becoming bankrupt, or taking any steps whatever towards entering into liquidation during the currency of any such Bill(s), and whether accepted conditionally or absolutely to sell all or any part of the goods forming the collateral security for the payment thereof at such times and in such manner as you or such Holders may deem fit, and after deducting usual commission and charges, to apply the net proceeds in payment of such Bill(s) with reexchange and charges the balance, if any, to be placed at your or their option against any other of our Bills, secured or otherwise, which may be in your or their hands, or any other debt or liability of mine/ours to you, or them, and subject, thereto, to be accounted for the proper parties. In case of loss at any time of goods insured, we authorize you, or the Holders thereof, to realize the policy or policies and charge the same commission on the proceeds as upon a sale of goods, and to apply the net proceeds, after such deductions as aforesaid, in the manner hereinbefore lastly provided.

7. In case the drawee of the Bill(s) request you or your correspondent on the date of maturity of the Bill(s) to postpone payment and if this is deemed reasonable by you or by your correspondent, no objection shall be raised by

八、茲雙方同意：倘押匯匯票因外來干預致不獲付款人承兌、不獲付款人或承兌人付款、或因當地法律規章或其他任何理由致使匯票無法付款，押匯款無從匯付　貴行時，不論該項匯票與（或）附屬單據是否退還，一經　貴行通知，我方願意立即償付匯票金額、利息及附隨之一切費用，　貴行如須增加擔保品，我方亦願意提供，絕無任何異議。

九、倘因匯票付款人、信用狀開證銀行、信用狀承兌行或信用狀保兌銀行無力償付債務，受破產宣告、查封、假扣押、假處分、拍賣等情事時、或因自請宣告破產或和解時，一經　貴行通知，我方願意償付　貴行匯票金額、利息以及附隨之一切費用。

十、如貨物變賣所得價款淨額不足以償付上開匯票所載金額（包括當時匯兌市價折合之損耗），茲授權　貴行「或　貴行之任何經理、代理人」或票據持有人，對於不足之款，得向我方發出匯票取償，但不影響該不足之數向其他背書人之追索權。茲同意：凡　貴行或票據持有人所出之帳單，即為變賣貨物已經受有損失之憑證，我方於該項匯票提示時，當即如數照付。

十一、不論變賣貨物之情事將否發生，茲授權「　貴行或　貴行之任何經理、代理人」或票據持有人，均得於匯票到期之前，接受付款人或

me/us to you or your correspondent's agreeing to it without notification to me/us.

8. We hereby agree that, should the Bill or Bills be not accepted by the drawees or not paid by the Drawees or Acceptors by intervention, or should it happen, that the Bill or Bills are not paid or the proceeds thereof are not transferred to you because of the local laws or regulations or for any other reasons, we shall pay the amount of the Bill or Bills with interest and other incidental charges incurred as soon as you inform us in this connection by cable or by mail, notwithstanding no return of the Bill or Bills and/or documents. Should you demand any additional security of us at same time, it shall be given by us without any objection.

9. Should the drawees of our Bill or Bills or the issuing, accepting or confirming banks of the relative Letter of Credit become insolvent, or bankrupt, be seized, provisionally seized, provisionally disposed of, or offered for auction, or even should the drawees or the above banks apply for bankruptcy or settlement by composition, we agree to pay you upon your notice the total amount of our Bill or Bills with interest and other additional charges.

10. In case the net proceeds of such goods shall be insufficient to pay the amount of any such Bill(s) with reexchanges and charges, we authorize you, or any of your Managers, or Agents, or the Holders for the time being of such Bill(s) as the case may be, to draw on us for the deficiency, without prejudice nevertheless to any claim against any endorser(s) of the said Bill(s) for recovery of same or any deficiency on the same; and we engage to honour such Drafts on presentation, it being understood that the Account current rendered by you or by such Holders shall be sufficient proof of sale and loss.

11. We further authorize you, or any of your Managers, or Agents, or the Holders for the time being of any such Bill(s) as aforesaid, whether the

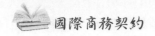

承兌人付款之要求，並於付款後將提單及其他貨運單據等，交與付款人或承兌人，倘　貴行或票據持有人准其提前支付時，「並得按照票據支付地之通常利率，計算折扣。」

十二、倘係承兌後交付貨運單據之匯票，我方授權　貴行，將附隨該匯票作為擔保品之貨運單據，於承兌人承兌該匯票後得交與承兌人，在此情形之下，倘因該匯票到期而承兌人不予付款，則凡因此而發生之後果，均由我方負其責任。我方當將該票所欠之全部款項、或一部分款項，及因此而增加之匯票及手續費如數償還　貴行，並擔保　貴行不因此而受任何損害。

十三、倘匯票付款人拒絕承兌或付款、或匯票到期前擔保貨品業已運抵目的港口，立書人授權　貴行或　貴行之通匯行辦理該匯票擔保品之卸貨、報關、存倉、保險等，　貴行或　貴行之通匯行認為維護此等貨品必要之任何措施，辦理上項措施所發生之有關費用，以及卸貨、報關、存倉及保險等各從業人員之過失，或因戰爭、天災或其他不可抗力因素所引起之任何損害，悉歸我方負擔。

十四、我方授權　貴行或　貴行之通匯行，以　貴行或　貴行之通匯行認為適合之任何方法寄送押匯匯票與（或）附屬單據。

十五、倘押匯匯票與（或）附屬單據在寄送中毀損或遺失，或視為已經毀損或遺失、或因誤送等意外情事，致令遲延寄達付款地時，得不必

aforesaid Power of Sale shall or shall not have arisen, at any time before the maturity of any such Bill(s), to accept payment from the Drawees or Acceptors thereof, if requested so to do, and on payment to deliver the Bill(s) of Lading and Shipping Documents to such Drawees or Acceptors; and, in that event, you or the Holders of any such Bill(s) are to allow a discount thereon, at the customary rate of rebate in the place where such Bill(s) are payable.

12.In case of D/A Bills we authorize you to deliver the documents to the acceptors against their acceptance of the Bill(s) drawn on them. In such a case, We undertake to hold you harmless from any consequence that may arise by your so doing and to pay you the amount or any balance of the Bill with reexchange and charges if the acceptors should make any default in payment at maturity.

13.Should the drawee of the Bill or Bills reject acceptance or payment of the said Bill or Bills, or should the collateral Goods arrive before the date of maturity of such Bill or Bills, we authorize your bank or your correspondent to unload, clear, warehouse the Goods, effect insurance thereon and do any and all other acts which your Bank or your Correspondent may deem necessary for the proper maintenance of the said Goods. In these cases, not only the expenses and cost incurred in the course of the above acts, but also any damage caused by those people or parties who deal with the unloading, clearance, warehousing and insurance in good or bad faith or by reason of war, natural disasters or any other Act of God shall be paid by us.

14.We authorize your bank or your correspondent to send the Bills and/or Documents to the place of payment by any method as you or your correspondent deems fit.

15.Should Bills and/or Documents be destroyed or lost in transit, or assumed as such, or their arrival at the place of payment is much delayed by accident

經任何法律手續，一經　貴行通知，我方願意根據　貴行帳簿之記載，作成新押匯匯票，倘可能者連同新附屬單據提供與　貴行；或隨　貴行之指示，立即償付　貴行匯票金額，以及附隨之一切費用。

十六、我方同意　貴行得就我方所有財產包括存於　貴行及分支機構、或　貴行所管轄範圍內之保證金及存款餘額等，均任憑　貴行移作共同擔保品，以清償任何現已發生或日後發生經已到期或尚未到期之任何未清償債務。

十七、在匯票或其他任何單據上所簽蓋之我方簽章或所寫文字，　貴行如認為與預先存驗於　貴行者相符、或與我方曾經使用於以前匯票或其他單據者相同時，即使其係偽造或被盜用，我方仍願負責，並償付　貴行因此而蒙受之損害。

十八、有關匯票與（或）擔保貨物之訴訟，以我方申請押匯之　貴行營業所在地地方法院為管轄法院。

十九、我方願遵守國際商會所刊布「押匯信用證統一慣例」並視其為本書之一部分。

二十、茲更經雙方協議同意：凡　貴行所有對於票據上，因退票而發生之一切權利，不因將擔保品交付與　貴行而受任何影響，亦不因　貴行行使票據權利上求償權而影響　貴行對我方所欠款項範圍以內在擔保品上占有之物權；此外關於我方店號、行莊公司，因股東、合夥人之死亡、退夥或加入新夥、或隨時而發生之其他人事變動，換言之，不論本處名稱、牌號及內部組織之如何變更，凡在我方繼續營業之時，本書所授權限及其設定，當繼續有效。凡每次我方匯票

such as mistransportation, a new Bill, and if possible, new Documents shall be presented to your bank by us according to your record book, at your demand without any legal procedures, or alternatively, at your option, the amount of the Bills, with all expenses, shall be paid to you by us.

16. We agree that all our property including securities and deposit balances which may now or hereafter be in your or your branches' possession or otherwise subject to your control shall be deemed to be collateral security for the payment of any indebtedness and liability now existing or hereafter created or incurred by us to you due or not due.

17. We shall be responsible for our signature, seal or writing used on the Bill(s) or any other documents accepted by you even though the signature, seal or writing is a forged or stolen one; in case you have concluded the same to be identical with those submitted to you beforehand or those used on a previous Bill or another document. Any damages, sustained by you therefrom, shall be paid for by us.

18. The jurisdiction of a judicial court regarding any legal action on my/our Bills and/or collateral goods shall be executed at the District Court at the location of your office where I/We submitted such Bills for negotiation or purchase.

19. We will observe the "Uniform Customs and Practice for Documentary Credits" fixed by the International Chamber of Commerce, and deem it as a part of this Letter.

20. Lastly, it is mutually agreed that the delivery of such collateral securities to you shall not prejudice your rights on any of such Bills in case of dishonour, nor shall any recourse taken thereon affect your title to such securities to the extent of my/our liability to you as above, and that, notwithstanding any alteration by death, retirement, introduction of new partners or otherwise in the persons from time to time constituting my/our firm or the style of my/our firm under which the business at present carried on by me/us may be from

經　貴行承購或貼現,均應認為我方又將已經訂立之本書重新訂立。
茲又經雙方同意:凡因　貴行所僱用之居間人或拍賣行之違約行為
而發生之結果,　貴行對於我戶並不負任何責任。此據。

20××年　　月　　日

　　　　　　　　　　　　　　申請人 _____
　　　　　　　　　　　　　　　　　　（簽章）

time to time continued, this Letter and the powers and authorities hereby given are to hold good as the Agreement with you on the part of the firm as aforesaid and that each negotiation of a Bill or Bills hereunder is to be treated as a renewal by or on behalf of the firm as then existing of the terms of this Agreement. It is also agreed that you are not to be responsible for the default of any Broker or Auctioneer employed by you for any purpose.

Dated this _____ day of _____ , 20 _____

Applicant's _____
Signature

§10 投標及履約保證金之保證申請書 (Application for Bid Bond/ Performance Bond Guarantee)

按廠商向政府機關或公民營公司投標時，該等機構常要求廠商提供銀行或公司所提供之保證，該保證得以保證書或擔保信用狀為之。以保證書為之者，廠商須先向銀行提出「投標及履約保證金之保證申請書」(Application for Bid Bond/Performance Bond Guarantee)，該申請書通常均訂有下列事項：

1. 申請日期 (Date of Application)。
2. 申請人之名稱、地址 (Name and Address of Applicant)。
3. 保證額度 (Amount of Guarantee)。
4. 保證標的 (Contract on Performance to be Guaranteed)。
5. 保證申請書之簽名、蓋印及交付 (Provision Indicating that the Application is Signed, Sealed and Delivered)。
6. 保證書開發及到期日期 (Date of Guarantee to be Issued and Expired)。
7. 保證費 (Guarantee Fee)。
8. 申請人之責任 (Duties of Applicant)。

CONTRACT

1. Obj... his Contract
1.1. ...tomer shall order and the Executor shal'...
consu... der the Technical Assignment (A...
Contra... ...ges) of the performance of...
1.2. Per... ...ment.
Technic...

2. Obligatio... ...e Parties
2.1. The Cu... ...r shall be obliged:
a) to pa... ...he work perform...
Contra...
b) to pro... ...mely all nec...
Executor; ...o provide f...
c) if necess... ...be oblige...
2.2. The Executor... ...under th...
a) to perform... ...e of wor...
b) upon perfo... ...n the ful...
c) to perform... ...ntract a...
determined in th... ...ntract a...

3. Procedure of Work
3.1. The Executor shall... ...m work t... ...any third...
3.2. The Executor may et... ...onditions o...
however, subject to terms... ...pproval or not...
...d for in this Contract... ...the Report on the performed work and... ...the period 1...
...or required. ...ion of work and within the... ...the Report... an... ...the date of receipt of the Report an...
...the Customer for signing. ...days of the performed work shall be de...
...tion, the performed work shall be... ...e assignment and incl...
...be argumented. In suc...

投標及履約保證金之保證申請書

致：××銀行

敬啟者：

申請人請求依所附格式及下列條件開發不可撤回之保證書：

1.為××公司（地址：××××）之利益開發。

2.依××公司名義（地址：××××）開發。

3.額度：＿＿＿＿＿＿＿＿＿＿＿＿＿＿＿＿＿＿＿。

4.為××契約（　　號碼）或要約之投標或履約保證金。

5.保證書開發日期：＿＿＿＿＿＿＿＿＿＿＿＿＿＿＿。

6.保證書到期日期：＿＿＿＿＿＿＿＿＿＿＿＿＿＿＿。

7.保證費：於提供保證後，給付保證金額之＿＿＿％。

申請人應負責償還　貴行自依本保證書給付之日起至申請人償還之日止，所給付之所有金額及依中華民國法令所許可而合計之利息。

APPLICATION FOR BID BOND/PERFORMANCE BOND GUARANTEE

To: ×× Bank

Dear Sirs:

The Undersigned hereby requests you to open an irrevocable letter of guarantee as per the attached form on the following terms and conditions:

1. In Favor of _____
(Name and Address)

2. For Account of _____
(Name and Address)

3. Amount (in words) _____

4. Representing bid bond/performance bond under invitation/contract no. _____

5. Date of Guarantee to be issued: _____

6. Expiration date of Guarantee: _____

7. Guarantee Fee: _____ % p.a. of amount given in the Letter of Guarantee, payable upon rendering the guarantee service as by you.

The Undersigned shall effect a full reimbursement to you for all the advances made by you under this Guarantee, with interest accrued according to the maximum interest rate permitted by the law and regulations of the Republic of China, prevailing at the time of the advance effected by you for the period from the date of such advance effected by you to the date of the performance by the Undersigned.

申請人就因本保證書所生爭議之費用（包括所有之法律費用），應負責償還之。

申請人於××年××月××日簽署且交付本文件。

大　安

_____

（簽名且／或蓋章）

The Undersigned shall be responsible for payment of all expenses arising from any and all disputes relating to the Guarantee, including all legal expenses.

IN WITNESS WHEREOF, the Undersigned has executed and delivered this application this _____

day of _____

Yours truly,

_____

Applicant Signature(s) and/or Seals

§11 本票 (Promissory Note)

所謂本票乃指發票人 (Maker) 簽發一定之金額，委託付款人 (Payer) 於指定之到期日，無條件支付與受款人 (Payee) 或執票人 (Holder) 之票據（參考票據法第三條）。依我國票據法第一二〇條之規定，本票均記載下列事項：

1. 發票日期 (Date of making)。
2. 發票地 (Place where made)。
3. 到期日 (Time payable)。
4. 無條件擔任支付 (Promise to pay)。
5. 受款人 (Payee)。
6. 金額 (Amount to be paid)。
7. 利息之支付 (Payment of interest)。
8. 付款地 (Place of payment)。
9. 表明本票之文字 (The nature of promissory note)。
10. 發票人簽名 (Signature of maker)。

CONTRACT

1. Obj...
1.1. ...tomer shall order and the Executor shall...
consu... ...der the Technical Assignment (A...
Contra... ...ges) of the performance of...
1.2. Per... ...nment.
Technic...

2. Obligatio... **...e Parties**
2.1. The Cu... ...shall be obliged...
a) to pa... ...he work perform...
b) to pro... ...mely all nece...
Executor; ...o provide f...
c) if necess... ...be oblige...
2.2. The Execut... ...e of wor...
a) to perform... ...under th...
b) upon perfo... ...n the ful...
c) to perform... ...ntract a...
determined in th...

3. **Procedure of Work**
3.1. The Executor shall... ...rm work...
...any third...
3.2. The Executor may e... ...conditions o...
however, subject to terms... ...pproval or not...
...led for in this Contract... ...required. ...tion of work and within the period in...
...e Report on the performed work an...
...e the Customer for signing. ...days of the date of receipt of the Report an...
...ion, the performed work shall be de...
...ssignment. In su...
...performed work shall be d...
...be argumented and incl...

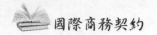

本　票

金額：　　　　　　　　　　　　　發票日：

　　　　　　　　　　　　　　　　發票地：

　　憑票無條件向×××銀行或其指定人於×××（到期日）支付該行×
××（金額）。本本票免除作成拒絕證書。付款無誤。

　　本本票以中華民國法為準據法。

　　　　　　　　　　　　　　　　發票人簽章：

利息：　　　　　　　　　　　　地址：

遲延利息：　　　　　　　　　　保證人簽章：

　　　　　　　　　　　　　　　　地址：

PROMISSORY NOTE

Amount: Date:

                                             Place:

I/We promise to pay unconditionally on demand to ×××　Bank or order the sum　×××　(amount) on　×××　(maturity date). The maker agrees to waive the protest procedure in the event of non-payment. Payable at　××× Bank.

This Note shall be governed by the laws of the Republic of China.

                                             Maker's Signature:

Interest: Address:

Past Due Interest: Guarantor's Signature:

                                             Address:

§12 分期付款契約 (Installment Payment Agreement)

　　於商業交易中，不論其為買賣或借貸，債權人為債務人給付價金或返還借款之便利，常允許債務人分期付款，當事人須就分期付款之條件予以約定，其主要須訂定下列事項：

1. 簽訂日期 (Date of Agreement)。
2. 當事人名稱及地址 (Names and Addresses of Parties)。
3. 債務之確認 (Recital of Debts)。
4. 清償之方式 (Payment of Debts, and Interests)。
5. 加速條款 (Acceleration Clause)。
6. 債務人之接受及簽名 (Debtor's Acceptance and Signature)。
7. 債務人抗辯之拋棄 (Debtor's Waiver of Defense)。

CONTRACT

1. Obj... this Contract
1.1. ...tomer shall order and the Executor shal...
consu... der the Technical Assignment (Ap...
Contra... ...ges) of the performance of...
1.2. Per... ...nment.
Technic...

2. Obligatio... e Parties
2.1. The Cu... ... shall be obliged:
...Contrac... mely all nec...
a) to pa... ... the work perform...
b) to pro... ...p provide t...
Executor; ...be oblige...
c) if necess... ...under th...
2.2. The Executor... ...e of wo...
a) to perform... ...n the ful...
b) upon perfor... ...ntract a...
c) to perform v... ...ntract a...
determined in th...

3. Procedure of Work
3.1. The Executor shall ... rm work ...
...however, subject to terms a... any third...
...ded for in this Contract... ...pproval or not...
not required. ...

...rm work and within the period 1...
...e Report on the performed work and...
...the Customer for signing...
...days of the date of receipt of the Report an...
...the performed work shall be d...
...tion, the performed work shall be d...
...shall be argumented and incl...
...the assignment. In such...

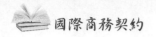

分期付款契約

20××年1月1日
××公司
中華民國臺北市
××街123號

×××先生／女士
有關：分期付款約定
　　本約定乃用以確認本人接受有關××金額之債務，以分期方式給付之。

　　敬請台端簽署本文件用為承認前述金額之債務，並同意以本約定之方式清償債務。

　　本人將接受自××年××月××日起，且於每月××日，由台端給付分期金額新臺幣××元之本金及年利率＿＿＿＿％之利息。

　　按期給付為本約定之重要考量因素，每次給付將先抵充利息，其次為本金。如未能按期給付，本人得請求即為給付餘額及利息。
　　如簽署人未能給付任何款項時，則簽署人就該交易未清償之金額及利息全部視為到期，且自該日起另計算利息。

INSTALLMENT PAYMENT AGREEMENT

January 1, 20 _____

××× Company

123 ××× Street

Taipei, Taiwan

R.O.C.

Dear _____

Re: Installment Payment Arrangement

　　This is to confirm the arrangement under which we will accept payment of our outstanding account of NTD _____ in installments.

　　You will sign and return the enclosed copy of this letter indicating admission of the full amount of the account and acceptance of the terms of our agreement.

　　We will, then, accept payment of the account, together with interest at the rate of _____ % per annum, calculated semi-annually not in advance, in consecutive, monthly installments of NTD _____ , commencing _____

(date)

and continuing on the _____ of each successive month until paid off in
　　　　　　　　　　(day)

full.

　　Time will be considered to be of the essence of this arrangement. Each payment will be applied, first, to accrued interest and, second, to principal.

　　If there is default in making any payment, at our option the full balance owing on the account, together with accrued agreed interest, shall immediately become due and payable and continue to accrue interest.

　　請於應給付首次分期款前簽署本約定書並返還之，否則，本約定書即為無效。

_____

（簽　名）

承認且接受。

　　簽署人於此承認前述之債務，且承認對他方無權主張抵銷或抗辯，並接受前述給付條件。

××年××月××日

_____

（簽　名）

Please return the signed copy of this agreement with your first payment before the commencement date of the monthly installments; otherwise this agreement is null and void.

Yours very truly,

————————————————

(signature)

Admission and Acceptance

The undersigned hereby admits the full amount of the above outstanding account and having no rights of set-off or counterclaim and accepts the above terms of payment.

Dated _____ 20 _____

————————————

(signature)

§13 附擔保之讓與契約 (Assignment with Warranties)

按債權人得將其債權讓與 (Assign) 第三人，但依債權之性質或特約，不得讓與或禁止扣押者，不在此限 (民法第二九四條)。債權之讓與如以出賣方式為之者，得經當事人約定，由讓與人負擔保責任 (民法第三五二條)。就讓與之方式，得以讓與書為之，其附加擔保 (Warranty) 者，則稱為附擔保之讓與契約 (Assignment with Warranties)，其通常載有下列條款：

1. 讓與之期日 (Date of Assignment)。
2. 當事人之名稱及地址 (Names and Addresses of Parties)。
3. 讓與之權利之說明 (Rights on Property Transferred)。
4. 讓與之對價 (Consideration for Transfer)。
5. 讓與人之擔保責任 (Assignment with Warranty of Underlying Contract)。
6. 讓與之接受 (Acceptance of Assignment)。
7. 受讓人解除契約之權利 (Right of Assignee to Terminate)。
8. 讓與人之承諾 (Covenant of Assurance by Assignor)。

CONTRACT

1. Obj
1.1. ~~...~~tomer shall order and the Executor shall
consu~~...~~der the Technical Assignment (Ap-
Contra~~...~~
1.2. Per~~...~~ges) of the performance of
Technic~~...~~nment.

...his Contract

2. Obligatio~~...~~e Parties
2.1. The Cu~~...~~ shall be obliged:
a) to pa~~...~~the work perform
Contract~~...~~mely all nece~~...~~
b) to pro~~...~~ provide f
Executor;~~...~~be oblige~~...~~
c) if necess~~...~~ under th
2.2. The Executor~~...~~e of wor~~...~~
a) to perform~~...~~ the ful~~...~~
b) upon perfor~~...~~
c) to perform~~...~~ntract a~~...~~
determined in th~~...~~ntract

3. Procedure of Work
3.1. The Executor shall~~...~~rm work t~~...~~
~~...~~ any third
3.2. The Executor may e~~...~~ conditions o~~...~~
however, subject to terms~~...~~pproval or not~~...~~
~~...~~led for in this Contract~~...~~
~~...~~t required.~~...~~tion of work and within the period in~~...~~
~~...~~r the Report on the performed work an~~...~~
~~...~~s the Customer for signing.~~...~~ the date of receipt of the Report an~~...~~
~~...~~ay) days~~...~~tion, the performed work shall be de~~...~~
~~...~~shall be argumented and inclu~~...~~
~~...~~he assignment. In such

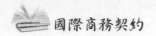

附擔保之讓與契約

　　茲承認已受領對價，×××（讓與人姓名，以下簡稱讓與人）乃讓與其於×××之權益予×××（受讓人姓名，以下簡稱受讓人）。

　　讓與人茲擔保其對讓與標的具所有權,且無須第三人同意得為本讓與。

　　於××年××月××日簽署並用印。

受讓人 _____　　　　　　　　讓與人 _____

ASSIGNMENT WITH WARRANTIES

For value received, which is acknowledged, _____ (the

(assignor name)

"Assignor") hereby assigns all interest owned in _____ to

_____ (the "Assignee").

(assignee name)

The Assignor also warrants to the Assignee that the Assignor owns the subject matter of this assignment and has the right to make this assignment without the consent of any third party.

Given under seal on _____

(date)

_____
The Assignee

_____
The Assignor

§14　貸款契約 (Loan Agreement)

貸款契約乃民法上之消費借貸契約，即當事人約定，由貸與人 (Lender) 移轉金錢之所有權於借用人 (Borrower)，而借用人負有返還義務之契約（民法第四七四條）。一般之借款契約包括下列事項：

1. 訂約日期 (Date of agreement)。
2. 當事人之姓名及地址 (Names and addresses of parties)。
3. 取得貸款意願之重述 (Recitals showing desire for the obtaining of a loan)。
4. 固定或不確定額度之借貸 (Flat amount or the amount as may be needed from time to time in the future)。
5. 貸款額度之限制 (Limitation on amount will be loaned)。
6. 取得貸款之證明 (Evidence of the loan as made)。
7. 貸款之利息 (Interest on loan)。
8. 貸款之返還 (Repayment of the sums loaned)。
9. 擔保之交付 (The giving of security)。
10. 借用人之擔保及陳述 (Warranties and representation of borrower)。
11. 財務報表之交付 (Furnishing of Financial Statements by borrower)。
12. 借用人之承諾 (Covenants of borrower)。
13. 借用人不履行之救濟 (Remedies in case of default by borrower)。
14. 借用人之帳簿 (Books and records of borrower)。
15. 擔保物之處分 (Disposition of proceeds of collateral)。
16. 服務費 (Service charges)。

17.借用人抗辯之拋棄 (Waiver by borrower)。

18.契約之修正 (Modification of agreement)。

19.契約之拘束力 (Binding effect of agreement)。

20.借用人營業及生命之保險 (Insurance on business and life of borrower)。

21.借用人其他債務負擔之限制 (Restrictions on other indebtedness of borrower)。

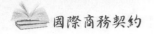

貸款契約

致：××銀行

　　立約人（以下簡稱借款人）茲因××銀行（以下簡稱該銀行）所予一次或多次貸款，信用狀之開發或資金之融通，茲同意下列各款：

第一條　擔保品定義：本約定書內所用「擔保品」一詞係指：(1)借款人在該銀行現在或將來各種貨幣之一切餘額，(2)為任何事由，不論其是否經同意，經送交或歸該銀行或其代理人所有保管或控制之下列各項款項，財產權利：(A)現金、票據、商業票據、期票債券、股票或其他證券、載貨證券、倉單、存款、保險單、各種債權及損害瞞哄權之請求權等；(B)上述之各項權益，所產生之利益，及代表上項權益之各種財產或其執行後所生之款項；及(C)借款人之其他利益、權利、財產、各種擔保品於送還該銀行或其代理人、合作人、商務關係人途中。或劃歸彼等，或為彼等持有時，將視同為該銀行所有。

第二條　本約定書內所稱「債務」一詞（不論一項或多項債務），係指借款人（或其所參加之合夥、財團、合營事業或其他團體機構，以下統稱借款人）對該銀行或該銀行現在或以後將有的利益關係之一切債務義務，及負債等，不論是項債務之形式為票據、證券、債券、匯票、或其他債務憑證；不論是項債務是否由借款、活期墊款、信用狀、透支、承諾費、遲延利息違約金、預先還違約金、

LOAN AGREEMENT

To: ×× Bank

The undersigned (hereinafter called the Borrower), in consideration of one or more loans, letters of credit or financial accommodations made, issued or extended by Bank (hereinafter called the Bank) hereby agree(s) as follows:

Section 1 The term "Security" as used herein shall mean: (1) all present and future credit balances in any currency of the Borrower with the Bank, and any other present or future Claim of the Borrower against the Bank, and (2) any of the following which have been or at any time shall be delivered to or otherwise into the possession, custody or control of the Bank or others acting in its behalf, for any purpose, whether or not accepted (A) money, negotiable instruments, commercial paper, notes, bonds, stocks or other securities, bills of lading, warehouse receipts, credits, insurance policies, choses in action, claims and demands; (B) any interest in, and any property represented by or called for in or which is the proceeds of, any of the foregoing, and (C) any other property, rights and interests of the Borrower. The Bank shall be deemed to have possession of any Security in transit to or set apart or held for it or any of its agents, associates or correspondents.

Section 2 The term "Liabilities" as used herein (whether in the plural or the singular) shall mean any and all indebtedness, obligations and liabilities of any kind of the Borrower (or of any partnership, syndicater association, joint venture or other group of which the Borrower is a member, hereinafter called collectively a partnership) to the Bank or in which the Bank shall have any interest, now or

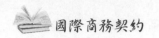

遲延還款違約金、契約、民事侵權行為、法律執行等所發生者；不論是項債務為本金、利息、承諾費、遲延利息違約金、期前還款違約金、以及遲延還款違約金所組成者；不論是項債務是既有或是或有，共同，或個別者，有擔保或無擔保者，屆期或未屆期者，直接或間接者，已清算或未清算者，或是否因借款人為主要連帶償還擔保人、背書人、保證人、信用票據連署人，或其他原因所引起者。

第三條　質權：借款人茲以所有擔保品向該銀行保證一切之債務，並將借款人對該擔保品之一切權利、名義、及利益，賦予該銀行享受繼續性之擔保權及抵銷等權。

第四條　不論債務是否即將屆期，該銀行得不通知；並除實際經收各項擔保品或其收益，及除本約定書另行明確規定者外，不因而發生任何義務；亦不因而解除借款人之任何義務，得隨時採取下列各項措施：

(1) 抵押債務人授權該銀行，對該擔保品行使各種表決權或其所有權利，而其效力與所有人所具有者相同；

(2) 將各項擔保品遷移至任何處所；

(3) 以該銀行或借款人名義：(A)請求提起訴訟，接受及取得因各項

hereafter existing, whether or not represented by notes, bonds, debentures; drafts or other evidences of indebtedness; whether arising out of loans, advances on open account, letters of credit, overdrafts, commitment fee, delayed interest payment, prepayment penalties, and delayed payments penalty, contract, tort or by operation of law or otherwise; whether consisting of principal, interest, commitment fee, delayed interest payment, prepayment penalties, and delayed payments penalty; whether absolute or contingent, joint or several, secured or unsecured, due or not due, direct or indirect, liquidated or unliquidated and whether incurred by the Borrower as principal surety, endorser, guarantor, accommodation party otherwise.

Section 3 As security for the Liabilities, the Borrower hereby pledges to the Bank all of the Security and hereby grants to the Bank a security interest in a general continuing lien upon and a right of set-off against, all right, title and interest of the Borrower in and to the Security.

Section 4 Whether or not any of the Liabilities shall be due, the Bank may from time to time, at its option, without notice, without thereby insurring any liability except to account for any Security or proceeds thereof actually received, and, except as otherwise expressly provided herein without thereby discharging any liability of the Borrower.

⑴ exercise all voting powers and all other powers with respect to the Security with the same force and effect as if the Bank were the absolute owner thereof; for this purpose the Borrower hereby authorizes the Bank on its behalf to exercise the voting powers or other powers with respect to the security;

⑵ remove any of the Security from any place to any other place;

⑶ in the name of the Bank or the Borrower; ⒜ demand, sue for,

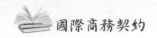

所應取得擔保品，或用交換該各項擔保品之已屆期之各項款
項、擔保品、或其他財產（包括本金、保險費、利息、紅利、
或其他收入、股息、認股權及各種貨物），或(B)同一部或全部
擔保品之債務人改組成或其他情事時，在該銀行認為滿意之條
件，和解或決定付款之日期或展期；修改有關擔保品之條件，
安排擔保品之分期付款；或為任何目的，將擔保品繳存於一委
員會或他人；

(4) 將因任何理由自擔保品所收到之淨得現在用以支付債務，或繼
續留作擔保品之一部；並將在借款人在該銀行結存之一部或全
部，或對該銀行之權利，用以支付債務；

(5) 繳回或放棄借款人擔保品之任何部分，或以任何擔保品交換借
款人所提供之其他擔保品；及

(6) 採取任何其他必要或適當之措施以保管或保存擔保品。

第五條　該銀行對於擔保品，除應依照法律規定範圍內依合理之注意加以
保管及保存外，無其他義務，並不負責通知或採取其他（借款人
所同意採取）之必要步驟，以保全對擔保品前關係人或前手之權
利，亦不對擔保品占有人所有之一切權利或義務加以行使，亦不
通告借款人是項權利或義務業經發生。

collect and receive any money, securities or other property (including principal, premium, interest, dividend or other income, stock dividends, rights to subscribe, and goods of all kinds) at any time due, payable or receivable on account of or in exchange for any of the Security, or (B) upon terms satisfactory to the Bank, whether upon a reorganization of the obligor on any instrument constituting part or all of the Security or otherwise, make any compromise or settlement with respect to or extend the time of payment of or otherwise amend the terms of any of the Security, arrange for the payment of any of the Security in installments, or deposit any of the Security with a committee or any other person for any purpose;

(4) apply toward the payment of the Liabilities; or continue to hold as part of the Security; any not cash receipts received for any reason from any of the Security, and apply toward the payment of the Liabilities any part or all of the balance of any account of the Borrower with or claim of the Borrower against the Bank;

(5) surrender or release any of the Security of the Borrower or exchange any of the Security for other security provided by the Borrower, and

(6) take any other action necessary or appropriate in connection with the custody or preservation of the Security.

Section 5 The Bank shall have no obligations with respect to the Security except to use reasonable care in the custody and preservation thereof to the extent required by law, provided however that the Bank shall not be obligated to give any notice or take any other steps necessary to preserve rights against any prior party or parties to any instrument (which steps the Borrower hereby agrees to take); nor to exercise any

第六條　一切構成擔保品之有形財產，不論其係由直接或用文件表示或證明者，借款人應隨時向該銀行所能接受可靠之保險公司投保全部火險及是項財產可能遭受之其他危險；並將保險單或該銀行核可之證件繳存該銀行；且於本約定書有效期間，在保險單內指明該銀行為保險金額之唯一受益人。一旦發生損失時，該銀行就其權益範圍內有優先受償之權。借款人如未能依照辦理，該銀行得（但無義務）維持是項保險，其一切費用將視同本約定書所稱之債務。

第七條　該銀行善意之認為借款人將來付款或履行義務之可能已受影響時，借款人經該銀行之要求，應隨時增繳該銀行所認為滿意之擔保品。

第八條　如

　　　　⑴借款人未能按期支付或償付依本約定書或與債務有關之任何合約所應付該銀行之任一宗本金債務者；

　　　　⑵借款人有依破產法聲請和解、聲請宣告破產、依法聲請公司重整，經票據交換所通知拒絕往來、或有停止營業、清理債務之情事者；

rights or obligations to the holder thereof nor to notify the Borrower that such rights, or obligations have arisen.

Section 6 With respect to all tangible property which, directly or through a document controlling or evidencing such tangible property, constitutes a part of the Security, the Borrower shall at its own expense at all times keep such tangible property fully insured against loss by fire and any other risks to which said property may be subject, with responsible companies acceptable to the Bank, and shall deposit with the Bank policies or certificates thereof in such form as the Bank shall approve, and, shall enter the Bank into the insurance policy as the only beneficiary during the validity of this Agreement. Loss, if any, payable under the said insurance policy shall be made to the Bank to the extent of its interest may appear. If the Borrower shall fail to do so, the Bank may, but shall not be obligated to, maintain such insurance and the expense thereof shall be deemed a Liability for all purposes of this Agreement.

Section 7 If at any time the Bank in good faith believes that the prospect of payment or performance by the Borrower is impaired, the Borrower shall, upon demand by the Bank, furnish such additional Security as is satisfactory to the Bank.

Section 8 If at any time

(1) the Borrower shall fail to repay of any Liabilities payable to the Bank when due under this Agreement or any other agreement in connection with the Liabilities;

(2) The Borrower shall be the subject of an application for composition (settlement) or a petition for declaration of bankruptcy under the operations or the Borrower shall make any arrangement for the settlement of its indebtedness;

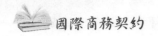

⑶ 借款人依其與該銀行之約定（不論係本約定書或與債務有關之任何合約）負有提供擔保之義務而不依約提供者或借款人未能按照第七條規定增補擔保品者；

⑷ （借款人為自然人時）借款人亡而其繼承人於繼承開始後聲明限定繼承或拋棄繼承者；

⑸ 該銀行獲悉借款人或其他代表人所提交該銀行之財務報表、與債務有關之合約、文件內容不實或有足以引起誤解之遺漏者；

⑹ 借款人未能按期支付依本約定書或與債務有關之合約、文件所應付該銀行之任一宗利息、費用或（本金除外之）其他應付款項，而未於該銀行給予通知後十五日內（下稱「補正期限」）補正者；

⑺ 擔保品或提供予該銀行作為債務擔保之其他擔保或保證變為無效、無法執行、價格低落不敷擔保債權，或任一保證人、擔保品提供者、或與擔保品有關之共同發票人、背書人、簽發人或其他相關之負責人發生第八條一至六款及七至十二款所列之各項情況，而未能於補正期限內補正者；

⑻ 借款人自該銀行所取得之融資，其實際資金用途與該銀行核定用途不符，而未能於補正期限內補正者；

(3) there is an obligation to provide security according to this agreement in connection with the Liabilities with the Bank and the Borrower shall fail to provide same or the Borrower shall fail to furnish Security as required by Section 7 hereof;

(4) (for individual Borrowers) the heirs or successors of the Borrower shall announce a limited inheritance or waive rights to inherit upon the Borrower's death or thereafter;

(5) the Bank shall find that any misrepresentation or misleading omissing has been made in any financial statement, Agreement or other document in connection with the Liabilities delivered to the Bank by or on behalf of the Borrower;

(6) the Borrower shall fail to pay interest, fees or any other sums (other than principal) of any Liabilities due to the Bank in accordance with this Agreement or any agreement in connection with the Liabilities, and such failure is not cured within fifteen (15) days after the Bank sends written notice to the Borrower ("Cure Period");

(7) the security or any other support or guarantee provided to the Bank for the Liabilities shall become invaild, unenforceable, reduced in value or otherwise become insufficient to secure the Liabilities or any of the events described in Section 8 (1)–(6), above or 8 (7)–(12),bellow shall occur with respect to any guarantor, provider of security or co-maker, indorser, issuer or any other person liable in any respect in connection with the security and any of such foregoing circumstance is not cured within the cure perido;

(8) the use of funds of any credit extended in connection with any Liabilities of the Borrower shall vary from the Bank's approved

(9) 借款人之財產或擔保品受強制執行、假扣押、假處分或其他保全處分致借款人對該銀行之債務有不能受償之虞，而未能於補正期限內補正者；

(10) 借款人未能按期支付其另與該銀行或第三者締結之其他合約下所應支付之款項，或借款人（不論係以主債務人或保證人身分）之金錢債務已發生加速到期或准許加速到期之情況，而未能於補正期限內補正者；

(11) 借款人違反其於本約定書或與債務有關之合約、文件下任一規定，而未補正者；或

(12) 借款人之管理、營運或財務狀況發生重大不利之變化，而未能於補正期限內補正者。

則一切債務毋須等待提出付款要求即應視為到期，借款人應予償付，此項原應提出付款要求之規定，借款人茲特聲明放棄，其後一切債務概將按照現行法定利率（如較當時流通之利率高）計息，但該銀行得於上述各項情況發生前或發生後，以書面通知放棄，中止，或修改各情事對債務之後果。每次之放棄，中止，或修改僅適用於各該特定情事。

purposes and such circumstances are not cured within the Cure Period;

(9) the property of the Borrower or the Security shall be the subject of compulsory execution, provisional attachment, provisional measures or other precautionary measures which is likely to adversely affect the Bank's recovery of the Liabilities and such circumstances are not cured within the Cure Period;

(10) the Borrower shall fail to make payment of any sums under any other Agreement (whether with the Bank or any third parties) when due or there occurs any event which accelerates or permits acceleration of the maturity of any indebtedness of the Borrower (whether to the Bank or any third parties) whether the Borrower is a primary obligor or guarantor and such circumstances are not cured within the cure period;

(11) the Borrower shall have violated its contractual obligations under this Agreement or under other agreement or instrument with the Bank such circumstances are not cured within the cure period; or

(12) there snall occur a material adverse change in the management, operations or financial condition of the Borrower and such circumstances are not cured within the cure period.

then in any such event, all Liabilities shall be due and payable forthwith without presentation or demand for payment, which are thereby expressly waived, and thereafter all Liabilities shall bear interest at the legal rate (if higher than the rate then applicable thereto), provided, however, that the Bank by notice in writing may waive, suspend or modify the effect of any such event upon any Liability either before or after the same shall have occurred. Each such waiver, suspension, or modification, shall apply only with

第九條　倘有本約定書第八條所列各項情事發生，該銀行得執行按照法律賦予受擔保之債權人對該擔保品之一切權利與救濟辦法，該銀行對擔保品之出售處分或其他必須通告之情事已於發生前五天，按照本約定書第十六條之規定，以書面通知借款人者，即可認為依法應送達之通知，已予照辦。

第十條　借款人應支付該銀行關於下列事項支出之一切合理費用（包括律師費及其他法律事項費用）：⑴各項債務或本約定書各項規定之執行；⑵各項擔保品之實際或試行變賣、交換、執行、催收、和解或清償；⑶擔保品之保管或保存；該銀行所付之各該項費用均視同本約定書主旨之債務，而得享受其一切利益。

第十一條　該銀行雖將擔保品全部變賣，或承受任何擔保品，以償付部分債務，借款人對於尚未償付之債務餘額（包括直至償付日止之利息）仍應繼續負責。

第十二條　本約定書所賦各項權利及救濟辦法為除其他原因發生及賦予之救濟辦法以外者，但法律禁止是項權利或救濟辦法之地區內不得行使之。該銀行之不採取行動或知情均不得被認為構成任何權力、權利或救濟辦法之放棄，單項或部分之行使不能妨礙其他權力、權利或救濟辦法之再加行使，該銀行對任何權力、權

respect to the specific instance involved.

Section 9 Upon any of the events specified in Section 8, above the Bank shall have and may exercise with respect to the Security all the rights and remedies given, allowed or permitted to a secured party by or under the law. Any requirement of reasonable notice imposed by law shall be deemed met if such notice is in writing and is mailed or delivered to the Borrower in the manner provided in Section 16 hereof at least five days prior to the sale, disposition or other event giving rise to such notice requirement.

Section 10 The Borrower shall pay all reasonable expenses (including lawyers' fees and other legal expenses) incurred by the Bank in connection with: (1) the enforcement of any of the provisions of this Agreement or of any of the Liabilities; (2) any actual or attempted sale, or any exchange, enforcement, collection, compromise or settlement of, any of the Security, or (3) the custody or preservation of the Security. Any such expense incurred by the Bank shall be deemed a Liability for all purposes of and shall be entitled to all of the benefits of this Agreement.

Section 11 Notwithstanding the realization by the Bank upon the entire Security, or the retention by the Bank of any Security in satisfaction of a portion of the Liabilities, the Borrower shall continue to be liable for any balance of the Liabilities (including interest to the date of payment) which shall thereafter remain unpaid.

Section 12 The rights and remedies given hereby are in addition to all other however arising, but it is not intended that any right or remedy be exercised in any jurisdiction in which such exercise would be prohibited by law. No action, failure to act or knowledge of the Bank shall be deemed to constitute a waiver of any power, right or

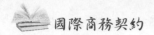

利或救濟辦法之行使或不行使，不負責任。

第十三條　借款人茲放棄票據之提示（除必要之承兌提示外），拒絕證書、
　　　　　拒絕證書之通知及對各項債務或擔保品所有票據不付款之通知
　　　　　（不論係初期到期、提前到期或其他情況者），及不論與是項票
　　　　　據有關或無關之其他一切通知或請求。

第十四條　除以書面協議，明確訂明其修正、修改或限制之事項，且經對
　　　　　其生效之一方簽署者外，本約定書不得修正、修改，或加以限
　　　　　制。本約定書將取代(1)借款人與該銀行間前所簽訂之結合貸款
　　　　　與債務擔保約定書；(2)取代及廢除以前及以後與該銀行所簽訂
　　　　　其他保管合約內不符合之約定。

第十五條　除書面另行協議並由該銀行逐次通知借款人變更者外，一切債
　　　　　務概應負擔至償還日止之利息，並於請求給付時，即向本行主
　　　　　要營業所或貴行指定之其他處所給付之，債務於星期日或公定
　　　　　假日到期者得於次日給付之,是項超出之日期之利息仍應計算。

第十六條　除於本約定書另行明確規定者外，各項增繳擔保品之要求，或

remedy hereunder, nor shall any single or partial exercise thereof preclude any further exercise thereof or the exercise of any other power, right or remedy. The Bank shall not be liable for exercising or failing to exercise any power, right or remedy.

Section 13 The Borrower hereby waives presentment (except presentment for acceptance when necessary), protest, notice of protest and notice of dishonor of any and all instruments included in the Liabilities or the Security, whether upon inception, maturity, acceleration of maturity, or otherwise, and any or all other notice and demand whatsoever, whether or not relating to such instruments.

Section 14 The Agreement shall not be amended, modified or limited except by a written agreement expressly setting forth the amendment, modification or limitation and signed by the party against which such amendment, modification to limitation is to be effective. This Agreement (1) shall supersede any General Loan and Collateral Agreement heretofore executed between the Borrower and the Bank and (2) shall supersede and override any inconsistent provisions of any custody agreement with the Bank, whether heretofore or hereafter executed.

Section 15 Unless otherwise agreed in writing and subject to alteration from time to time by the Bank, on notice to the Borrower, all Liabilities shall bear interest at rate of ＿＿＿ per annum to the date of payment and shall be payable at the principal place of business of the Bank or any place otherwise designated by the Bank, upon demand. Any Liability which matures on a Sunday or public holiday shall be payable on the next succeeding business day and such additional time shall be included in the computation of interest.

Section 16 Demands for additional Security and any other demands or notices

其他向借款人提出之要求或通知，如對借款人以電話，或書面專送郵寄或電報方法，送至下列地址，或借款人以書面通知該銀行之其他地址者（如借款人係一公司或合夥商號，則對其任何職員或合夥人），則是項要求或通知應認為已按照規定送達借款人。

第十七條　借款人應於每三個月向該銀行提供真實確實之報表一次。遇有資產大量出售，借款人在業務管理上、控制上，或政策上之重大變更，借款人受有不利之裁判，本約定書第八條所列各項情事之發生，應立即（如可能時，則於二十天前）通知該銀行，借款人應經常提供該銀行所要求之資料，並於經要求時，准許該銀行查閱其帳簿及記錄，並予以摘錄或抄錄。

第十八條　由銀行之轉讓，代理人：對於此項債務或擔保品，該銀行得依本約定書有所行為或為委任人或以參與人為之或為其代理人，並得允許他人參與或受讓上述債務及本約定書，並得將擔保品轉移於任何委任人、參與人或受讓人，在此種情形之下，則本約定書所用「銀行」一詞將包括所有之委任人、參與人及受讓人。而各委任人、參與人，及受讓人均得為經在本約定書所列名之人享受本約定書一切之權益。該銀行如仍保持債務或擔保品之一部，則將繼續享受應有之利益，但對於經轉移之擔保品則將解除對其一切之請求與責任，該銀行可不通知委任人、參與人或受讓人或徵得其同意，依其自己之決定行使本約定書所賦予之一切權利。該銀行之代理人或其所擔保之人均得如經在

to the Borrower shall, unless otherwise expressly provided herein, be deemed duly and properly made or given if made or given to the Borrower (or, if the Borrower is a corporation or partnership, to any officer or partner) by telephone, or in writing delivered by hand to or telegraphed or mailed by ordinary mail to the Borrower at the address indicated below or at such other address as the Borrower may furnish to the Bank in writing.

Section 17 The Borrower shall furnish to the Bank quarterly financial statements which shall in our respects be true and correct and shall give prompt notice (Twenty days in advance when possible) of any bulk sale of assets or any material change in the managements; control on business policies of the Borrower or of any judgment against the Borrower or of the occurrence of any event mentioned in Section 8 hereof. The Borrower shall from time to time provide any additional information requested by the Bank and, upon request of the Bank, shall permit inspection of the Borrower's books and records and the making of extracts and copies therefrom.

Section 18 Assignments by Bank; Agents. The Bank may act hereunder, or with respect to any of Liability or Security, on behalf of or as agent for any principal or participant, and may grant participations in or assign of any of the Liabilities, and may grant participations in or assign this Agreement and may transfer any security to any such principal, participant in or assignee of any the Liabilities. In any such case, the term "Bank" as used herein shall include all such principals, participants and assignees, each of whom shall have all the benefits of this Agreement as if named herein. The Bank shall continue to have the benefits hereof if it retains any interests in any of the Liabilities and Security but shall be fully discharged from all

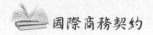

本約定列名之人享受本約定書之利益。

第十九條　本約定書係由一人以上簽署時，所有簽署人均應負連帶責任，所謂「借款人」一詞係指各該簽署人，而有關債務或擔保品之各項規定均適用於彼等一人或全體所有之債務或擔保品，本約定書對於各借款人之法定繼承人、法定代表人、繼承人、及受讓人均有約束效力。簽署人係合夥事業，則雖合夥事業有變更，本約定書仍將繼續對該合夥事業有效。

第二十條　本約定書適用於現在及將來之一切交易，不論交易性質是否為現時所考慮者，如借款人與該銀行之交易將一次或多次完成，本約定書對於其後完成之新交易仍為有效，該銀行對於本約定書之接受並不能認為係對放款或貸款有所承諾。

第二十一條　倘遇中華民國政府對立約定書人應付該行之利息要求扣繳稅款時，立約定書人同意如數補償之。倘遇中華民國政府對該行應付國外貸款機構之利息，要求扣繳稅款時，立約定書人亦同意補償之。

claims and responsibility for any security so transferred. The Bank may in its descretion exercise any of the rights herein granted without the consent of or notice to any principal, participant or assignee. Every agent or nominee of the Bank shall have the benefit of this Agreement as if named herein.

Section 19 If more than one person signs this Agreement: they shall be jointly and severally liable hereunder, the term "Borrower" shall refer to any one or more of such persons, and the provisions hereof regarding the Liabilities or Security shall apply to any of the Liabilities or Security of any or all of such persons. This Agreement shall be binding upon the heirs, legal representatives, successors and assigns of each Borrower. If the undersigned is a partnership, this Agreement shall continue in force notwithstanding any change in such partnership.

Section 20 This Agreement shall apply to all existing and future transactions, whether or not of the character contemplated at the date hereof. If all transactions between the Bank and the Borrower shall at any time or times be closed, this Agreement shall nonetheless be applicable to any new transactions thereafter entered into or effected. The acceptances of this Agreement shall not be deemed a commitment by the Bank to make any loan or extend any credit.

Section 21 The Borrower hereby agrees to compensate the Bank in full amount in the event the R.O.C. Government demands payment of withholding tax on the interest payable by the Borrower to the Bank. The Borrower further agrees to reimburse the Bank for withholding tax, if any, in the event the R.O.C. Government, demands payment of withholding tax on the interest payable by the Bank to its source of bonds abroad.

第二十二條　本約定書內所用之各標題，僅為方便之用，不能被解釋為對條文內容有所限制，本約定書之解釋將按照中華民國之法律規定為之，其簽訂各方之權利將遵照是項法律規定辦理。

** 借款人茲確認借款人於詳細閱讀上述約定書各條款內容並該銀行商議後，完全了解並同意上述約定書之各項規定，其並包括約定書第八條五、七、十、十一、十二款規定之加速條款事由。

西元 20　　年　　月　　日

借款人：＿＿＿＿＿＿＿＿＿　　貸款人：＿＿＿＿＿＿＿＿＿

地　址：＿＿＿＿＿＿＿＿＿　　地　址：＿＿＿＿＿＿＿＿＿

Section 22 The section headings herein are inserted solely for convenience and are not to be construed as limitations upon the text. This Agreement is to be construed according to and the rights of the parties hereunder are to be governed by the laws of the Republic of China. This document has been drawn up in English and Chinese. In the event of a dispute, this Agreement shall be interpreted on the basis of its English text. In the event of any litigation pertaining to this Agreement, the Borrower hereby submits and consents to the non-exclusive jurisdiction of the Taipei District Court.

**THE BORROWER HEREBY EXPRESSLY ACKNOWLEDGES THAT IT UNDERSTANDS AND SPECIFICALLY AGREES TO THE FOREGOING TERMS AND CONDITIONS, IN PARTICULAR, THE ACCELERATION CLAUSES SET OUT IN SECTION 8 (5), (7), (10), (11) AND (12), AFTER SEPARATELY REVIEWING AND NEGOTIATING SAME WITH THE BANK.

Sign on this _____ day of _____ 20 _____

Borrower: _____   Lender: _____
Address: _____   Address: _____

§15 聘僱契約 (Employment Agreement)

聘僱、聘用、受僱契約等，均指僱用人 (Employer) 與受僱人 (Employee) 就受僱人提供服務，而僱用人給付報酬之契約（參考民法第四八二條），該種契約之約定，應包括下列事項：

1. 簽約日期 (Date of agreement)。
2. 當事人之姓名及住址 (Names & addresses of parties)。
3. 受僱人之義務及責任 (Duties of employee)。
4. 聘僱之場所 (Place of employment)。
5. 受僱人之工作時間 (Time to be devoted)。
6. 僱傭之報酬 (Compensation of employee)。
7. 僱傭之期間 (Term of employee)。
8. 僱傭之終止 (Termination of employment)。
9. 契約之延長 (Extension of agreement)。
10. 工作費用之補償 (Reimbursement for expense)。
11. 契約之修正 (Modification of agreement)。
12. 受僱人之行為禁止或限制 (Restrictive covenants)。
13. 糾紛之解決 (Dispute settlement)。
14. 受僱人之假期 (Vacation of employee)。
15. 本契約之有關通知事項 (Manner of giving notices)。
16. 姓名之使用 (Use of name)。
17. 對他造財產（如專利）之歸屬 (Right of either party in property)。
18. 契約權義之讓與 (Assignment)。
19. 僱用人之保障 (Protection of employer)。
20. 契約違反之救濟 (Breach of contract)。
21. 僱用人之責任 (Duties of employer)。
22. 其他重要事項 (Other provisions)。

CONTRACT

...his Contract
1.1. ...tomer shall order and the Executor shall...
consu... ...der the Technical Assignment (Ap-
Contra... ...ges) of the performance of
1.2. Per... ...nment.
Technica...

...e Parties
2. Obligatio...
2.1. The Cu... ...shall be obliged:
Contract... a) to pa... ...he work perfor...
b) to pro... ...mely all nec...
Executor; ... ...provide...
c) if necess... ...be oblige... ...under th...
2.2. The Executo... ...e of wor...
a) to perform... ...n the ful...
b) upon perfo... ...ntract a...
c) to perform... ...determined in th...

3. Procedure of Work ...m work...
3.1. The Executor shall... ...any third...
3.2. The Executor may en... ...conditions o...
however, subject to terms... ...pproval or not...
...ded for in this Contract... but required... ...ion of work and within the period in...
...the Report on the performed work and... ...the Customer for signing. ...days of the date of receipt of the Report... ...ion, the performed work shall be argumented and incl...
...th... assignment. In such...

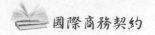

聘僱契約

僱用人信封之公司名稱

受僱人之姓名及地址
致 A 先生：
據我們昨日之會談，本公司將提供下列之要約：

本公司將聘請台端為某州某市之商店經理人，台端將依董事會之限定有其權利及責任。

為台端所給付之服務或經要求之服務，本公司將每年以分期之方式，及每一月之最後之營業日支付台端薪水（元）。

台端之僱傭任期（年），且於台端提供之服務日起開始，至二〇＿＿＿年＿＿＿月＿＿＿日。
於本契約之期間，台端將盡最大之努力及全部之時間以謀取本公司之利益，且應盡台端之能力以履行本公司所要求之義務，台端且於每一營業日之商店之經營期間，應於本公司之工作場所為工作，且應履行本公司所指示、指導或控制等所要求之工作。

本契約將由以台端與本公司，且對台端與本公司、台端之繼承人與受讓人有其拘束力。

GENERAL EMPLOYMENT CONTRACT

(Letterhead of Employer)

(Name and Address of Employee)

Dear Mr. ___A___ :

In accordance with our conversation of yesterday, we will make you the following offer:

We will employ you as manager of our store at No. _____ Street in the City of _____, State of which such powers and duties in that connections may be fixed by our Board of Directors (or by us).

For the services rendered by you, as hereinafter required, we will pay to you a salary $ _____ per year in equal installments on the last business day of each month (or on Friday of each week).

The term of your employment shall _____ years from the date on which you are to begin your services, to wit, the _____ day of _____, 20 _____ .

During the term of this agreement, you shall devote your best efforts and entire time to advance our interests, shall perform the duties reguired of you to the best of your ability, and shall be at our place of business hereinbefore mentioned on each business day during the hours that we require the store to be kept open and shall perform such other work as may be required of you by us under and subject to our instruction, direction and control.

This agreement shall inure to the benefit of and shall be binding upon you and us, and your and our successors and assigns.

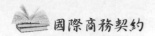

　　如本要約為台端所接受，請於本契約所載之空白地方簽名，且將該信函退還於本公司。

　　　　　　　　　　　　　　　　　　　　　　　　　催用人之簽名

接受且認可
受僱人之全名

If this offer is acceptable to you, please sign a copy of this letter in the space provided below for your signature and return it to us.

(Signature of Employer)

Accepted and approved

(Signature of employee)

§16 專業人員聘僱契約 (Retainer Agreement for Professionals)

契約往往聘請專任顧問提供服務，該契約具有委任與僱傭之特質，依其性質而定，其主要之條款如下：

1. 服務 (Services)。
2. 報酬 (Compensation)。
3. 費用 (Costs)。
4. 陳述 (Representations)。

CONTRACT

this Contract

...tomer shall order and the Executor shal'
...der the Technical Assignment (Ap

1. Obj
1.1. The
consu
Contra ...ages) of the performance of
1.2. Per
Technica ...nment.

he Parties

...er shall be obliged:
...the work perfor

2. Obligatio
2.1. The Cu
a) to pa ...mely all nece
Contrac
b) to pro ...o provide f
Executor, ...be oblige
c) if necess ...e under th
2.2. The Executo ...e of wor
a) to perform ...n the ful
b) upon perfo ...ntract a
c) to perform ...ntract a
determined in th

3. Procedure of Work
3.1. The Executor shall ...rm work ...
...any third
however, subject to terms ...conditions o
3.2. The Executor may e ...pproval or noti
...ed for in this Contract
...not required.

...on of work and within the period i...
...the Report on the performed work a...
...the Customer for signing.
...days of the date of receipt of the Report a...
...on, the performed work shall be...
... shall be argumented and incl...
...ssignment. In suc...

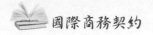

專業人員聘僱契約

本人茲代表 A 聘僱 B 為顧問，以提供法律協助、諮商及 A 之任何工作。

為提供該等服務，A 同意支付年度顧問費新臺幣＿＿＿＿＿＿＿＿，而該等金額得使 A 享有＿＿＿小時之服務，如該服務之時數超過此基本時數時，A 同意以每小時新臺幣＿＿＿＿＿＿＿＿＿＿之金額給付予 B。

A 茲同意給付 B 所有必要之費用。

A 茲承認 B 有關所進行之事務不作保證或承諾，而就 B 所為之陳述成為該案件未來之預估，僅為 B 之意見。

A 茲認可於本契約簽署時，收到該契約。

本契約於＿＿＿年＿＿＿月＿＿＿日於臺灣臺北簽署之。

律師　　　　　　　　　客戶
B　　　　　　　　　　A

＿＿＿＿＿＿＿＿＿＿　　＿＿＿＿＿＿＿＿＿＿

RETAINER AGREEMENT

I, the undersigned, on behalf of A (hereinafter "A"), hereby retain and employ the law firm of _____ , _____

_____ , Taipei, Taiwan, R.O.C. (hereinafter referred to as "B"), to provide legal assistance, consultation, and representation for the subsidiaries and affiliates of A in Taiwan as requested.

As compensation for professional services, A agrees to pay Attorney a non-refundable retainer fee of _____ New Taiwan Dollars (NT$ _____), which will entitle A to _____ hours of B's services. If any hours in excess of those covered by the retainer fee are required to perform the services set out above, or to perform other services as directed by A, A agrees to pay B at the rate of NT$ _____ per hour for attorney services.

A agrees to pay any and all reasonable and necessary expenses incurred by B on its behalf or necessary to advance its cause.

A acknowledges that B makes no guarantees or promises with regard to the matters undertaken by B, and that any statements which B makes relative to a successful outcome are B's opinion only.

A acknowledges receipt of a copy of this Agreement concurrently with the execution hereof.

Entered into in Taipei, Taiwan, R.O.C. on _____ , 20 _____ .

Attorney   Client

B   A

By: _____   By: _____

§17 短期人員僱傭契約 (Short-term Employment Agreement)

按僱傭契約除長期或正式僱傭關係之外，商業實務上越來越多企業採行所謂之短期僱用契約，此等契約仍為僱傭契約，僅為就相關之僱傭條件有所變更，其主要之條件如下：

1. 僱傭之期間 (Term)。
2. 試用期間 (Probation Period)。
3. 受僱人之義務 (Employee Duties)。
4. 報酬 (Remuneration)。
5. 勞工保險 (Labor Insurance)。
6. 終止 (Termination)。
7. 準據法 (Governing Law)。

CONTRACT

1. Ob... ...his **Contract**
1.1. ...tomer shall order and the Executor shall
consu... ...der the Technical Assignment (Ap...
Contra... ...ges) of the performance of...
Technica... ...nment.
1.2. Per...

...**e Parties**
2. **Obligatio...** ...shall be obliged...
2.1. The Cu... ...T shall be obliged...
    a) to pa... ...the work perform...
    Contra... ...mely all nec...
    b) to pro... ...o provide f...
    Executor; ...be oblige...
    c) if necess... ...under th...
2.2. The Executo... ...e of wo...
    a) to perform... ...n the ful...
    b) upon perfo... ...m work t...
    c) to perform v... ...ntract a...
    determined in th...

3. **Procedure of Work**
3.1. The Executor shall p... ...m work t... ...any third...
3.2. The Executor may e... ...conditions o...
however, subject to terms a... ...pproval or not...
...led for in this Contract... ...not required... ...tion of work and within the period in... ...work an...
...e the Report on the performed work an... ...the Customer for signing. ...the Customer for signing. ...) days of the date of receipt of the Report an... ...) days of the date of receipt of the Report... ...) days... ...ion, the performed work shall be d... ...ion, the performed work shall be argumented and incl... ...be assignment. In such...

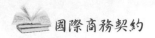

短期人員僱傭契約

A 公司（簡稱 "A"）及 B 受僱人（簡稱 "B"），於 ＿＿＿＿ 年 ＿＿＿＿ 月 ＿＿＿＿ 日簽訂本契約。

A 茲同意提供 B 暫時之短期之僱用，而 B 同意接受此等短期之僱用。

依此，當事人同意 B 接受依以下條款之僱用：

1. 受僱期間

    當事人同意僱用期間自簽訂之日起至 ＿＿＿＿ 年 ＿＿＿＿ 月 ＿＿＿＿ 日，共 ＿＿＿＿ 日／月。於本契約屆期後，本契約即自動終止。

2. 短期僱用及試用期間

    B 之認可此僱用乃短期，而 B 之同意此契約以一定期間為試用期間之短期契約，該試用期間為 ＿＿＿＿ 天，如 A 認為 B 未能履行其義務時，A 得隨時於試用期間內終止該契約。

3. B 之義務

    B 應履行下列義務：

_____

_____

_____ 。

    B 同意於 A 指示之時間、地點於期間內履行其義務。

4. 費用

    A 應給付 B 每月新臺幣 ＿＿＿＿ 元，因 B 乃短期之僱用，B 同意其不得

SHORT-TERM EMPLOYMENT AGREEMENT

THIS AGREEMENT entered into on the _____ day of _____, 20 _____ between _____ ("A") and _____ ("B").

WHEREAS, A is willing to offer B short-term employment, and B is willing to accept such short-term employment.

NOW THEREFORE the parties agree that B shall accept such employment subject to the following terms and conditions:

1. Term of Employment

The parties agree that the term of this agreement shall be _____ days/months commencing from the date of this agreement to the _____ day of _____, 20 _____ . At the expiration of the term of this agreement, the employment relationship shall automatically terminate without any conditions.

2. Short-Term Employment and Probation Period

B acknowledges that the employment offered by A is short-term in nature and agrees that he or she is willing to accept appointment as a short-term employee of A subject to an initial probation period of _____ days. In the event that A considers that B cannot perform his or her assigned duties, A has the right to terminate this agreement at any time during the probationary period.

3. B's Duties

B's duties shall include _____

_____

_____

_____ .

B shall perform such duties during the hours and at the time and place directed by A.

4. Remuneration

A shall pay B a salary of NT$ _____ per month. Because the employment

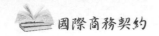

享有如永久僱用之權利，包括紅利及退休福利等。

5. 勞工保險

　　如 A 之保險人同意且依法律許可，A 茲同意給予 B 與 A 之永久受僱人相同條件之勞工保險。

6. B 之責任

　　於本契約有效期間，B 應依其主管之指示履行其義務，應與善良管理人之注意義務且就其工作保持秘密。如 B 違反其契約，應對 A 負損害賠償責任，A 得依其裁量向 B 請求或終止本契約，A 亦保留其權利對 B 起訴為求償，或追訴刑事責任。

7. 終止

　　如因 B 之過失而致下列情形發生者，A 得於本契約有效期間內不經公司而終止本契約：

⑴B 於簽訂本契約為不實陳述，而使 A 受有損害，

⑵B 對 A 之員工、家屬及代表人、共同工作之同仁有暴力、重大過失之侮辱，

⑶B 依法院之最終判決受拘役或較重之刑罰，且該處分並無緩刑或罰金之情形，

⑷B 違反該等契約或 A 之工作規則，屬情節重大者，

⑸B 故意使 A 之機器設備、工具、原料或產品發生損害，或將該技術或

is short-term, B agrees that he or she shall not be entitled to receive benefits which permanent employees receive, including bonuses and retirement benefits.

5. Labor Insurance

If permitted by law and accepted by A's insurer, A agrees to insure B subject to the same terms and conditions as A's permanent employees.

6. B's Obligations

During the term of this agreement, B shall comply with the directions and instructions of his or her supervisors, perform his or her duties with due diligence, and keep all matters relating to A's business strictly confidential. In the event that B breaches any obligations specified in this paragraph and such breach results in damage to A, A may in its discretion reprimand B or terminate this agreement. A also reserves the right to institute suit against B for compensation or to impose criminal liability.

7. Termination

In case of any one of the following conditions which is due to the worker's fault, A may terminate this agreement during its effective term without advance notice:

1) Where employee makes a false statement at the time of signing the agreement, thereby misleading A and making A liable to suffer damage,

2) Where B acts violently against or hurls insult at members of A, their families, A's representative or fellow workers,

3) Where B is sentenced by a court as final judgment to detention or a heavier punishment, and the sentence has not been commuted to a probation or fine,

4) Where B violates this agreement or working rules and the case is considered to be serious in nature,

5) Where B purposefully inflicts damage to or excessively abuses machinery,

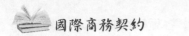 國際商務契約

商業機密洩露於他人，導致 A 受有損害，

(6) B 於未經請假或無正當理由連續休假未工作達三日，或於一個月中有六日未工作者。

8.準據法

因本契約所生之爭議，應適用中華民國法律，臺北地方法院有其專屬管轄權。

當事人同意於上述日期簽訂本契約。

當事人

A 公司：   B 受僱人：

_____代表人_____   _____

職稱：　_____

equipment, tools, raw materials, products or other article belonging to A, or intentionally discloses the technological or business secrets of A causing a loss to A,

6) Where B absents himself from work for three consecutive days without justifiable reasons, or for six days in a month.

8. Governing Law

Any dispute or controversy between the parties with respect to this agreement shall be determined in accordance with the law of the Republic of China. The District Court of Taipei shall have exclusive jurisdiction.

IN WITNESS WHEREOF, the parties have entered into this agreement on the date first above indicated.

A   B

By _____   _____
Title _____

§18　授權書 (Power of Attorney)

按外國人於臺灣申請外人投資 (FIA)，均必須向經濟部投審會申請，而該事務之工作由代理人為之，較為具彈性與便利，故通常須有此等授權書，具有代理之特質，其主要之條款如下：

1. 委任人 (Appointer)。
2. 代理人 (Agent)。
3. 代理人之職權 (Capacity of the Agent)。
4. 期間 (Term)。

CONTRACT

1. Obj...
1.1. ...tomer shall order and the Executor shall...
consu... der the Technical Assignment (A...
Contra... ...ges) of the performance of...
1.2. Per... nment.
Technica...

2. Obligatio... e Parties
2.1. The Cu... r shall be obliged...
a) to p... he work perfor...
Contract... mely all nec...
b) to pro... provide...
Executor; ... be oblige...
c) if necess... under th...
2.2. The Executo... e of wor...
a) to perform... n the ful...
b) upon perfo... ntract a...
c) to perform... ntract a...
determined in t... ntract a...

3. Procedure of Work
3.1. The Executor shall... m work...
3.2. The Executor may e... any third...
however, subject to terms... conditions o...
...ed for in this Contract... pproval or not...
...required. ...on of work and within the period...
...on of work... the Report on the performed work an...
...the Customer for signing.
...days of the date of receipt of the Report a...
...ion, the performed work shall be...
...signment. In su...

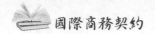

授權書

致：中華民國經濟部投資審議委員會

　　A公司乃依香港法律而設立，此公司所在地為＿＿＿＿，該公司擬任命住所於＿＿＿＿之＿＿＿＿先生為在臺之代理人，其有完全之權限指派其他代理人或撤銷其指派，並得代理本人依中華民國法律為＿＿＿＿業務，提出申請、修正申請及／或溝通以向主管機關取得核准之執照。

　　本授權書於＿＿＿＿年＿＿＿＿月＿＿＿＿日經＿＿＿＿先生因合法授權而為簽署並為公司用印。

　　　　A公司

　　　　＿＿＿＿（代表人）＿＿＿＿＿＿＿＿＿＿＿＿＿＿＿＿
　　　　（姓名）
　　　　＿＿＿＿（職位）＿＿＿＿＿＿＿＿＿＿＿＿＿＿＿＿

POWER OF ATTORNEY

TO THE INVESTMENT COMMISSION, MINISTRY OF ECONOMIC AFFAIRS REPUBLIC OF CHINA

We, A Corporation, a corporation organized and existing under the laws of Hong Kong and having its principal office at _____ hereby constitute and appoint Mr. _____ of _____, situated at _____ Taipei, Taiwan, the Republic of China to be our attorney in the Republic of China, with full power to appoint agents to act as attorney on our behalf and to revoke the appointment of such agents, to file and prosecute on our behalf an application for approval of all required licenses to invest in and operate a _____ enterprise in the Republic of China under the laws and regulations relating to such enterprises, to make corrections, amendments and/or supplementations therein, to conduct all procedures concerning same, to receive service on our behalf of all documents and communications relating to these matters and generally to represent us and to transact all business on our behalf in regard to these matters.

IN WITNESS WHEREOF, A Corporation has caused this instrument to be executed by _____ thereunto duly authorized, and its corporate seal to be hereunto affixed, on the day of _____, 20 _____.

A CORPORATION

By: _____

Name (please print): _____

Title: _____

§19 指派書 (Letter of Appointment)

外國公司於臺灣必須指派本國人或外國人自然人為代表人，以參加股東會或董事會，其具有代理之特質，然亦可能必須簽訂委任契約，其主要之條款如下：

1. 委任人 (Appointer)。
2. 代理人 (Agent) 或代表人 (Representative)。
3. 委任之事務 (Capacity)。
4. 撤回 (Revocation)。
5. 期間 (Term)。

CONTRACT

1. Obj... [t]his Contract

1.1. [The Cus]tomer shall order and the Executor shal'...
consu... [char]ges) of the performance of
Contra... der the Technical Assignment (Ap...
1.2. Per... nment.
Technic...

2. Obligatio... [th]e Parties

2.1. The Cu...r shall be obliged:
Contrac... the work perform...
a) to pa...
b) to pro... mely all nece...
Executor;... provide f...
c) if necess... be oblige...
2.2. The Executo... under th...
a) to perform... e of wor...
b) upon perfo... n the ful...
c) to perform v... ntract a...
determined in th... ntract a...

3. Procedure of Work

3.1. The Executor shall ... m work t... any third ...
3.2. The Executor may et... conditions o...
however, subject to terms ... pproval or not...
...ed for in this Contract ...
...not required. ...tion of work and within the period 1...
...he Report on the performed work and...
...the Customer for signing. ...days of the date of receipt of the Report an...
...tion, the performed work shall be dec...
...n shall be argumented and incl...
...t shall be ...

指派書

　　A 公司為依日本法設立而主事務所為＿＿＿＿東京，日本之公司，茲為設立 A 公司於臺灣，乃指派自然人＿＿＿＿（護照號碼為＿＿＿＿之人）為代表人，該代表人同意以 A 公司之董事會或常務董事協商，以進行該代表之事務。

　　除 A 公司以書面撤回者外，本委任書所賦予之授權應繼續有效。

　　本契約由當事人於 2000 年＿＿＿＿簽訂之。

代表人　　　　　　　　　　　　A 公司代表人

_____　　　_____

　　　　　　　　　　　　　　　_____

　　　　　　　　　　　　　　　職位

LETTER OF APPOINTMENT

A Company, a corporation organized and existing under the laws of Japan, and having its principal office located at _____, Tokyo, Japan desiring to nominate a natural person to act as its representative to establish A Taiwan Ltd., in the Republic of China, hereby appoints _____, Passport Country and No. _____ of _____ as such representative _____ agrees to consult with the board of directors or the managing director of A Company, prior to taking any action as such representative.

The authority given in this appointment shall continue in force until revoked by an authorized representative of A Company, by written notice.

IN WITNESS WHEREOF, the parties have caused this agreement to be executed on _____, 20 _____.

	By, for and on behalf of
REPRESENTATIVE	A Company
by: _____	by: _____
_____	_____
Name (Please Print)	Name (Please Print)
	_____
	Title

§20 保證 (Guaranty Agreement)

所謂之保證 (Guaranty) 乃指當事人約定，保證人 (Guarantor) 於他造之債務人不履行債務時，由其代為履行責任之契約（參考民法第七三九條）。保證契約一般均約定下列事項：

1. 訂約之日期 (Date of agreement)。
2. 當事人之姓名及地址 (Names and addresses of parties)。
3. 擔保債務性質之說明 (Recitals showing nature of debt or contract to be paid or performed by the original debtor)。
4. 保證之約因 (Consideration of guaranty)。
5. 保證之本質 (Nature of guaranty)。
6. 保證之受益人 (For whose benefit guaranty made)。
7. 保證之拘束力 (Binding effect of guaranty)。
8. 保證之期間 (Length of time guaranty is to continue)。
9. 保證人責任之限制 (Limitations on liability of guarantor)。
10. 保證人終止保證之權利 (Right of guarantor to terminate guaranty)。
11. 債務之擔保 (Security for payment)。
12. 保證人抗辯權之拋棄 (Waiver by guarantor)。
13. 通知 (Notice)。
14. 準據法 (Governing law)。
15. 糾紛解決 (Dispute settlement)。

CONTRACT

1. Obj[...]his Contract
1.1. [...]tomer shall order and the Executor shall
consu[...]der the Technical Assignment (Ap-
Contra[...]ges) of the performance of
1.2. Per[...]nment.
Technic[...]

2. Obligatio[...]e Parties
2.1. The Cu[...]r shall be obliged:
a) to pa[...]mely all nec[...]
Contrac[...]
b) to pro[...]o provide f[...]
Executor;[...]be obliged[...]
c) if necess[...]under th[...]
2.2. The Executo[...]e of wor[...]
a) to perform[...]n the ful[...]
b) upon perfo[...]ntract a[...]
c) to perform [...]ntract a[...]
determined in th[...]

3. Procedure of Work
3.1. The Executor shall [...]m work [...]any third [...]
3.2. The Executor may et.[...]conditions o[...]
[...]wever, subject to terms [...]pproval or not[...]
[...]ed for in this Contract [...]
[...]st required.[...] on of work and within the period in[...]e the Report on the performed work an[...]e the Customer for signing.
[...]days of the date of receipt of the Report a[...]ion, the performed work shall be d[...]shall be argumented and incl[...]ssignment. In su[...]

保　證

致：××銀行

敬啟者：

　　鑒於　貴行向＿＿＿＿（以下稱「客戶」）以有擔保或無擔保方式提供財務融通或客戶因此項融資對　貴行負有債務（財務融通一詞包括但不限於貸款、信用及融資之授與、應收帳款及一切商業票據之貼現及買入及擔保品之設定或其他授信或契約行為等），立保證書人茲不可撤銷並無條件連帶保證屆期（於約定日到期、提前到期或其他情事到期）償付並履行客戶應付　貴行之一切債務，不論其發生之原因或時間為何，包括但不限於下列債務（以下合稱「保證債務」）：⑴貴行現已或將來提供他人之款項或基於客戶票據而現已或將來提供他人之款項，⑵因　貴行提供客戶款項或與客戶往來所生之一切直接或間接債務，⑶客戶於　貴行帳戶項下現已或將來積欠　貴行之款項或客戶現已或將來直接或間接積欠　貴行之債務，不論該債務到期已否，不論其係單獨或與他人共同負清償責任，不論其係以主債務人或介入地位負擔債務，亦不論其為確定或或有之債務，⑷所有與上開款項、債務或其擔保品有關之利息、佣金、成本、手續費及費用，包括律師費用及催收費用，立保證書人並不可撤銷並無條件同意於客戶未如期清償保證債務之全部或一部時（包括債務到期日屆至，提前到期或其他情事到期），應不待　貴行之通知或請求立即依照客戶於保證債務下所規定之清償方式、貨幣種類及清償地點清償保證債務。除法律之強行規定外，前述保證責任之有效存續不受⑴表彰或與任何財務融通有關之任何文件或安排（以下稱「融資文件」）之合法性、真實性、有效性、拘束力或執行力之有無，或該文件之任何更改、修正、追加或捨棄之影響，⑵　貴行未為請求、實行或執行該融資文件，⑶客戶享有抗辯、抵銷或扣抵之權利或其他法律上免責事由或保證人得主張抗辯事由存在而受影響。

GUARANTY

To: ×× Bank

Dear Sir:

In consideration of your extending financial accommodation with or without security to or for account of _____ (hereinafter referred to as the "Customer") or in respect of which the Customer may be liable in any capacity (the term financial accommodation including, without limitation, extension of loans, credit or accommodation, or discount or purchase of, or loans on, accounts, leases, instruments, securities, documents, chattel paper and other security arrangements, or other property, or entering into exchange contracts), each of the undersigned guarantor(s) hereby absolutely, irrevocably and unconditionally guarantees to you, jointly and severally with the Customer, the due and punctual payment and discharge when due (whether at stated maturity, by acceleration or otherwise) of all the obligations and liabilities (collectively hereafter referred to as the "Guaranteed Obligations") of the Customer to you thereon or thereunder howsoever or whensoever incurred including without limitation: (1) the repayment of all moneys advanced or which may be advanced by you to the Customer or to others on the faith of the paper of the Customer; (2) all liabilities direct or indirect to which you may become subject as a result of making advances to or dealing with the Customer; (3) payment of all moneys which are now or shall at any time or from time to time hereafter become due or owing from the Customer to you on the general balance of account or for which the Customer now is or shall at any time hereafter become liable to you either directly or indirectly, whether matured or not, whether alone or jointly with others, and whether as principal or surety and whether absolute or contingent; and (4) all interest, commissions, costs, charges and expenses including all attorney's fees, costs and expenses of collection which may be

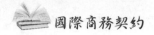

立保證書人茲聲明並同意下列條款:

第一條　本證書屬繼續性之保證，其效力及拘束力不受下列之影響:⑴客
　　　　戶結束在　貴行之帳戶，⑵　貴行於任何時候受到任何給付，⑶
　　　　任何帳目經予結清，⑷其他情事。本保證書之效力及　貴行於本
　　　　保證書下之權利不受⑴客戶破產、公司重整、清算或其他程序，
　　　　⑵立保證書人中之一人或數人之保證中斷，或⑶立保證書人或任

incurred in respect of such advances or liabilities or any securities therefor. In addition, each of the undersigned guarantor(s) hereby irrevocably and unconditionally agrees that if the Customer shall default in the payment when due, upon maturity, acceleration or otherwise, of all or any of the Guaranteed Obligations, each of the undersigned guarantor(s) will forthwith pay the same, without any notice or demand, at the place, in the funds and currency and in the manner required of the Customer with respect to such Guaranteed Obligations. The obligations of the undersigned guarantor(s) under this guaranty shall, subject to any mandatory provision of law, remain operative and in full force and effect, irrespective of: (1) the regularity, genuineness, validity, legality, binding effect or enforceability of, or any change in or amendment, supplement or waiver with respect to or modification of, any instrument, writing, or arrangement evidencing or otherwise relating to or the subject of any financial accommodation (each such instrument, writing or arrangement as now or hereafter in effect being hereafter referred to as, and included in the term, "Credit Arrangement"); or (2) the absence of any action or attempt to make a claim under or realized upon or enforced the same; or (3) the existence of any defense, set-off or counterclaim which the Customer or any other person may now or hereafter have or any circumstances which might otherwise constitute the legal or equitable discharge or defense of a guarantor.

Each of the undersigned guarantor(s) hereby declares and agrees that:

Section 1. This guaranty shall be a continuing guaranty and shall be operative and binding notwithstanding that (1) at any time or times the Customer's account with you may be closed or (2) any payments from time to time may be made to you or (3) any settlements of account may be effected or (4) any other thing whatsoever may be done, suffered or permitted. The validity and enforceability of this

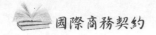

何於本保證書下負載之人中一人或數人死亡、喪失或限制行為能力之影響。

第二條　本保證書任何條款或規定之修改或免除，應以書面為之並經當事人簽名，否則對　貴行不具拘束力。

第三條　貴行得隨時依其認為適當之方式另行接受其他保證或收取擔保品債務或其部分，或交還、免除、捨棄或以其他方式處分該保證或擔保品及　貴行現所持有之保證或擔保品。但　貴行於本保證書下所享有之權利不受影響。在法令允許最大範圍內，　貴行得在不免除立保證書人之責任下，拋棄、變更或不執行任何保證或擔保品，或與立保證書人達成債務和解。

第四條　在法令允許最大範圍內，　貴行得不經通知立保證書人或取得立保證書人之同意，隨時允許客戶、保證債務之債務人或表彰保證債務融資文件上之債務人延期清償或履行，或以　貴行認為適當之方式與債務人達成債務和解，但　貴行於本保證書下所享有之權利不受影響。

guaranty shall not be affected and your rights hereunder shall not be prejudiced by (1) the bankruptcy, reorganization, liquidation or other proceeding of the Customer or (2) the discountinuance of this guaranty as to one or more of the undersigned guarantor(s) or (3) the death or loss or diminution of capacity of any of the undersigned guarantor(s) or any party responsible hereunder.

Section 2. No alteration or waiver of this guaranty or of any of its terms, provisions or conditions shall be binding on you unless made in writing and signed by the party or parties to be charged.

Section 3. You shall be at liberty (without in any way prejudicing or affecting your rights hereunder) from time to time to take such further or other guaranty or guaranties or security for the Guaranteed Obligations or any part thereof as you may deem proper, and/or release, discharge, abandon or otherwise deal with any such guaranty or guaranties or security or securities or any part thereof or with any guaranty or security (or any part thereof) now held by you, all as you may consider expedient or appropriate. To the greatest extent permitted by law, you may, without exonerating the undersigned guarantor(s), give up, modify or abstain from perfecting or taking advantage of any guaranties or securities and accept or make any compositions or arrangements.

Section 4. To the greatest extent permitted by law, you may from time to time, without notice to or further consent of the undersigned guarantor(s), grant to the Customer or to any person or persons liable to you for or in respect of the Guaranteed Obligations, or any part thereof, or in respect of any Credit Arrangement evidencing the same, or any part thereof, any indulgence whether as to time, payment, performance or otherwise and may compound with all or any of such persons as you

第五條　保證人償付　貴行保證債務之款項，應依民法第三二一條、第三二二條及第三二三條之規定抵償保證債務。

第六條　立保證書人同意於客戶現在或將來積欠　貴行債務（包括利息、成本及費用）全部清償前，立保證書人不得向客戶或就客戶或任何其他保證人之財產行使或主張權利或收受清償（依本保證書規定清償而代位承受　貴行之權利者亦然）。

第七條　貴行任何授權職員所簽發載明保證債務之書面文件，除有明顯錯誤外，對立保證書人具有確定之拘束力，立保證書人茲放棄一切就　貴行與客戶或保證債務之債務人現在或將來往來之方式及就　貴行目前或以後持有之任何保證或擔保品及該擔保品項下之任何貨品或財物提起異議之權利。　貴行勿須經先客戶或他人或就　貴行所持有之保證或擔保品追索求償，即可向立保證書人逕行求償。

shall see fit, all without in any way prejudicing or affecting any of your rights hereunder.

Section 5. You may apply any monies received from the undersigned guarantor(s) under this Guaranty to any portion of the Guaranteed Obligations in accordance with Articles 321, 322 and 323 of the Civil Code of the Republic of China.

Section 6. Each of the undersigned guarantor(s) agrees that until after final payment in full of all sums (including interest, costs and expenses) which may be or become payable by the Customer to you at any time or from time to time, the undersigned guarantor(s) will not exercise, claim or receive the benefits of any right (by subrogation to your rights, contribution or otherwise, arising by virtue of any payment by any of the undersigned guarantor(s) pursuant to any provision of this guaranty) to any payment by the Customer or by any other guarantor(s) out of the property of the Customer or any other guarantor(s).

Section 7. The statement in writing of your authorized officers of the amount of any Guaranteed Obligation shall be binding upon and conclusive against the undersigned guarantor(s) absent of any manifest error in computation and all rights to question in any way your present or future method of dealing with the Customer or any dealing with any person or persons now or hereafter liable to you for the Guaranteed Obligations or any part thereof or with any guaranties or securities now or hereafter held by you or with any goods or property covered by such securities or any of them are hereby waived. You shall not bound to exhaust your recourse or take any action against the Customer or other person or persons or the guaranties or securities you may hold before requiring or being entitled to payment from the

第八條　於客戶或其保證人發生破產、公司重整、和解、解散或分配財產之情事時，　貴行於本保證書下所享有之權利不因　貴行未申報債權或未申報全部債權而受影響，　貴行得自行決定申報債權與否，但皆不免除、減少或影響立保證書人對　貴行所負之責任。如保證債務經客戶清償而嗣後又因客戶有破產、公司重整、和解或其他類似程序（不論其係由客戶自行提起或由他人提起）情事而向　貴行取回或　貴行返還時，本保證書所規定之保證應繼續適用於保證債務，該保證債務自始為未受清償。

第九條　立保證書人於法令許可最大範圍內，茲放棄民法債編第二十四節「保證」所規定之各項權利，包括但不限於先訴抗辯權，立保證書人並放棄一切提示、請求、作成拒絕證書及通知，包括但不限於債務不履行之通知、拒絕證書之通知、拒絕付款之通知，同意本保證書之通知，及保證債務發生或存在之通知。如本保證書由二人以上簽署時，立保證書人間應負連帶責任，其效力並及於立保證書人之繼承人、遺囑執行人、遺產管理人、法定代理人、繼受人及受讓人。就任一立保證書人保證責任之免除或就保證債務之債務人所為債務之免除，並不影響其他立保證書人之保證責任。

undersigned guarantor(s).

Section 8. Upon the bankruptcy, reorganization, composition, winding up or other distribution of assets of the Customer or of any surety or guarantor for any indebtedness of the Customer to you, your rights hereunder shall not be affected or impaired by your omission to prove your claim or to prove your full claim. You may prove such claim or may refrain from proving any claim as you see fit without in any way releasing, reducing or otherwise affecting the liability to you of any of the undersigned guarantor(s) hereunder. Furthermore, if any Guaranteed Obligations is satisfied by the Customer but is subsequently recovered from or repaid by you, in whole or in part, in any bankruptcy, reorganization, composition or other similar proceedings instituted by or against the Customer or otherwise, the guaranty set forth herein shall continue to be fully applicable to such Guaranteed Obligations to the same extent as though the payment so recovered or repaid has never been made.

Section 9. Each of the undersigned guarantor(s), to the greatest extent permitted by law, waives all the rights as provided for in the various articles of Section 24 of the Book of Obligations of the Civil Code of the Republic of China including, without limitation the right of ordinis beneficium and waives all presentments, demands for performance, protests and notices, including without limitation notices of non-performance, notices of protest, notices of dishonor, notices of acceptance of this guaranty and notices of the existence, creation or incurring of any and all indebtedness. If this guaranty is executed by more than one party, then each obligation hereunder shall be jointly and severally binding on them and each of them, their and each of their heirs, executors, administrators, other legal representatives,

第十條　依本保證書規定為清償財務融通之款項，立保證書人應依各該融
　　　　資文件所規定之清償方式、地點及貨幣種類清償各該保證債務。
　　　　依本保證書規定所應付之款項，不得抵銷、扣抵或扣繳任何稅捐
　　　　或規費（「稅捐」），立保證書人同意代　貴行繳納任何稅捐或款
　　　　項，以使　貴行佳賓收取依本保證書下所規定之應付款項，立保
　　　　證書人並同意經　貴行請求，立即償付　貴行一切已繳納或應繳
　　　　之稅捐，決不使　貴行蒙受任何損失。立保證書人並同意於遲延
　　　　給付本保證書下任何應付款項時，應向　貴行支付自到期日起至
　　　　實際清償日止依決定最高利率計算之利息，立保證書人並同意於
　　　　遲延給付或履行本保證書下之債務時，應給付　貴行一切催收之
　　　　各項合理成本及費用（包括但不限於律師費用及墊款等）。

successors and assigns. Any one or more of the parties executed this guarantee, or any other party liable upon or in respect of any of the Guaranteed Obligations, may be released without affecting the liability of any of the parties not so released.

Section 10. All payments made hereunder in respect of any financial accommodation shall be made to you in the currency required to satisfy the applicable Guaranteed Obligations, at the place and in the manner specified in the relevant Credit Arrangement. All payments hereunder shall, be made to you without set-off or counterclaim and free and clear of and without deduction and withholding for or an account of all present, and future taxes, levies, duties, fees or withholdings whatsoever, if any, now or hereafter imposed ("Taxes"). Each of the undersigned guarantor(s) will pay on your behalf the full amount of all Taxes as may be so imposed or levied and such additional amounts as may be necessary so that the net payment received by you after payment of all such Taxes shall be not less than the amounts provided for hereunder. Each of the undersigned guarantor(s) will indemnify you, hold you harmless against and reimburse you, upon demand, for any Taxes paid or payable by you. Further, in the event that any of the undersigned guarantor(s) shall default in the payment in full when due of any sum payable hereunder, each of the undersigned guarantor(s) agrees to pay interest on the sum not so paid in full when due from the due date thereof until the date the same is paid in full at the highest rate then permitted by applicable law. In addition, each of the undersigned guarantor(s) agrees to pay all reasonable costs and expenses of collection (including, without limitation, legal fees and disbursements of counsel) in case of

第十一條　所有依本保證書規定所應付之款項，應依本保證書規定之清償
　　　　　地（「指定地」）及指定種類之貨幣（「指定貨幣」）給付之，立
　　　　　保證書人依指定貨幣給付之義務，不因以其他種類貨幣之給付
　　　　　或依任一以其他種類貨幣決定受領給付而免除或消滅，但　貴
　　　　　行因該給付或該判決而依該債務所規定之清償地及指定貨幣全
　　　　　額受領該債務及本保證書下金額之清償者，不在此限。立保證
　　　　　書人並同意　貴行實際受領之金額少於依本保證書規定應付指
　　　　　定貨幣之金額時，上述以指定貨幣給付款項之義務得另行以訴
　　　　　請求之，不受就本保證書其他款項已取得勝訴判決而影響，立
　　　　　保證書人同意取得一切必要之政府核准（包括中央銀行之核
　　　　　准），俾為前述之給付。

第十二條　在保證債務及立保證書人於本保證書下之債務完全清償以前，
　　　　　非經　貴行事先書面同意，立保證書人不得出售、轉讓、設定
　　　　　抵押或負擔或以其他方式處分其現有或將來取得之不動產。

default occurs in the payment or performance of any obligation of any of the undersigned guarantor(s) hereunder.

Section 11. All payments hereunder shall be made to you in the currency (the "Agreed Currency"), at the place (the "Agreed Place") provided herein. The Obligation of the undersigned guarantor(s) to make payment in the Agreed Currency of any amounts due hereunder to you shall not be discharged or satisfied by any tender, or any recovery pursuant to any judgment which is expressed in or converted into any currency other than the Agreed Currency, except to the extent such tender or recovery shall result in the actual receipt by you at the Agreed Place of the full amount of the Agreed Currency expressed to be payable in respect of such liability and all other amounts due hereunder. Each of the undersigned guarantor(s) agrees that the obligation to make payments in the Agreed Currency as aforesaid shall be enforceable as an alternative or additional cause of action for the purpose of recovery in the Agreed Currency of the amount (if any) by which such actual receipt shall fall short of the full amount of the Agreed Currency expressed to be payable in respect of any amount due hereunder, and shall not be affected by judgment being obtained for other sums due under this guaranty. Each of the undersigned guarantor(s) agrees to obtain all necessary government approvals, including the approval of the Central Bank of China, to effectuate the foregoing.

Section 12. Until the Guaranteed Obligations and all obligations of the undersigned guarantor(s) hereunder shall have been paid in full, each of the undersigned guarantor(s) shall not, without your prior written consent, sell, transfer, create mortgage or encumbrance on or otherwise dispose of any of the properties of the undersigned

第十三條　本保證書為客戶對　貴行所有現在已發生及將來發生之保證債
　　　　　務（包括經由持續交易而隨時展期續作或到期還款再更新而發
　　　　　生之各項保證債務）之連續保證。保證人得隨時撤回其對現尚
　　　　　未發生之保證債務之保證責任，但該項撤回僅於　貴行確實收
　　　　　訖該項撤回之書面通知時，始生效力，且該項撤回對　貴行收
　　　　　到該撤回通知前已發生之保證債務，就連帶保證人之義務及
　　　　　貴行之權利而言，皆無任何影響。

第十四條　本保證書中任一部分或條款之無效或執行力，並不影響本保證
　　　　　書其他部分或條款之效力或執行力。　貴行就本保證書下權利
　　　　　之放棄，非經　貴行以書面為之不生效力，且　貴行就特定事
　　　　　項所為權利之放棄，其效力不及於　貴行為該權利後所生之相
　　　　　同事項或其他違約事項，但　貴行在該權利放棄文件有明示之
　　　　　相反記載者，不在此限。

第十五條　於保證債務及本保證書項下之債務完全清償以前，立保證書人
　　　　　應於　貴行要求時提交　貴行所有關於其財務狀況之資料。立
　　　　　保證書人應於　貴行要求時責成其各債權人提出書面陳述，確
　　　　　認並無任何得主張撤銷或否認本保證效力之法律原因存在。

guarantor(s).

Section 13. This guaranty is a continuing guaranty for any and all present or future Guaranteed Obligations, including Guaranteed Obligations arising under successive transactions which shall either continue such Guaranteed Obligations or from time to time renew such Guaranteed Obligations after such Guaranteed Obligations have been satisfied. This Guaranty may be revoked at any time by any of the undersigned guarantor(s) in respect of future Guaranteed Obligations which are not already existing. Such revocation shall be effective upon your actual receipt of written notice of revocation from the undersigned guarantor(s) and shall not affect any of the undersigned guarantor's obligations or your rights with respect to Guaranteed Obligations which come into existence prior to your actual receipt of such notice.

Section 14. The invalidity or unenforceability of any part or provision of this guaranty shall not affect the validity or enforceability of any other part or provision of this guaranty. Further, you shall not be deemed to have waived any of your rights under the guaranty, unless you shall have signed such waiver in writing. No such waiver, unless expressly so stated therein, shall be effective as to any transaction which occurs subsequent to the date of such waiver nor as to any continuance of a breach after such waiver.

Section 15. Until the Guaranteed Obligations and all obligations of the undersigned guarantor(s) hereunder shall have been paid in full, each of the undersigned guarantor(s) shall, upon your request, cause their respective creditors to confirm, in writing, that there exist no legal grounds for invalidation or contestation of the binding effect of the guaranty.

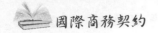

第十六條　本保證書應依中華民國法律解釋。

第十七條　因保證書事項涉訟時，立保證書人同意以臺灣臺北地方法院為非專屬管轄法院。

第十八條　立保證書人茲都有並保證：(1)立保證書人有能力及權限簽署並履行本保證，如保證人為一公司，則已具備所有獲得公司授權之必要程序，以簽署及執行本保證合約；(2)本保證書構成立保證書人合法、有效及有拘束力之債務，得依其規定條款執行；(3)立保證書人簽訂或履行本保證書，並不違反任何法律規章，亦不構成任何立保證書人簽訂其他合約或文件或對立保證書人或其財產有拘束力之合約或文件下之違約事項或需取得他造當事人同意之事項；(4)立保證書人已取得為簽訂或履行本保證書依法取得之政府核准、同意、授權或登記。

第十九條　本保證書以中、英文訂立，如有文義兩歧，則如在中華民國提起訴訟者，以中文本為準，如在中華民國以外地區提起者，以英文本為準。

Section 16. This guaranty shall be construed in accordance with the laws of the Republic of China.

Section 17. In the event of any litigation pertaining to this guaranty, each of the undersigned guarantor(s) hereby submits and consents to the nonexclusive jurisdiction of the Taipei District Court.

Section 18. Each of the undersigned guarantor(s) represents and warrants that (1) the undersigned guarantor(s) has full capacity and right to make and perform this guaranty and in the case where the undersigned guarantor is a company, the corporate undersigned guarantor has taken all necessary corporte action to authorize execution and performance of this guaranty; (2) this guaranty constitutes legal, valid and binding obligation of the undersigned guarantor(s) enforceable in accordance with its terms; (3) the making and performance of this guaranty do not and will not violate the provisions of any applicable law or regulation or order, and do not and will not result in the breach of, or constitute a default under or require any consent under, any agreement, instrument or document to which the undersigned guarantor(s) is a party or by which the undersigned or any of the property of the undersigned may be bound or affected; and (4) all consent, approvals, licenses and authorizations of, and filings and registrations with, any governmental authority required under applicable law and regulations for the making and performance of this guaranty have been obtained or made and are in full force and effect.

Section 19. This guaranty is executed in both English and Chinese versions. In the event of any discrepancy in meaning between the English and Chinese texts, the Chinese version shall govern if a lawsuit to settle the dispute arising herefrom is initiated in the R.O.C., and the

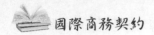

　　各保證人茲確認其於詳閱本保證書並與　貴行商議條款內容後，完全了解並同意於本保證書下之義務係就客戶之上述保證債務負連帶清償之連帶保證責任，並謹此明示同意第三條、第四條、第七條及第九條所列之各項同意及權利拋棄條款，為證明起見，保證人茲於相關人條款上簽名蓋章為憑。

　　　　　　　　　　中華民國　　年　　月　　日
　　　　　　　　　　　　連帶保證人：
　　　　　　　　　　　　身分證號碼：
　　　　　　　　　　　　地　　　址：

銀行同意簽章：
XX 銀行

_____

English version shall govern if a lawsuit is initiated outside the R.O.C..

EACH OF THE UNDERSIGNED GUARANTOR(S) BY INITIALLING/ CHOPPING OF CLAUSES 3, 4, 7 AND 9 ABOVE, EXPRESSLY ACKNOWLEDGES THAT ITS/HIS/HER GUARANTEED OBLIGATIONS UNDER THIS GUARANTY ARE JOINT AND SEVERAL WITH THE CUSTOMER AND SPECIFICALLY AGREES TO THE WAIVERS AND CONSENTS SET OUT THEREIN AFTER SEPARATELY REVIEWING AND NEGOTIATING SAME WITH THE BANK.

Dated this _____ day of _____, 20 _____

Guarantor

By _____

Name:

ID Card No.:

Address:

_____ :

×× Bank

_____

§21 合資契約 (Joint Venture Agreement)

投資或合資時，一般均先簽訂合資契約 (Joint Venture Agreement) 或股東協議書 (Shareholders Agreement)，於合資事宜有共識時，始進行公司、合夥或其他法人之設立。該合資契約通常須就下列事項為協商及約定：

1. 簽約日期 (Date of Agreement)。
2. 當事人名稱及地址 (Names and Addresses of Parties)。
3. 合資事業之業務 (Business to be Conducted)。
4. 出售股份之限制 (Restrictions on Sale of Stock)。
5. 表決權之約定 (Voting Agreement)。
6. 公司員工之聘僱 (Employment of Corporate Officer)。
7. 當事人之出資 (Contributions of Parties)。
8. 出資之給付 (Payment of Contributions)。
9. 企業之管理 (Management of Enterprises)。
10. 董事及經理人之待遇 (Compensation of Director Manager)。
11. 利潤之分配 (Division of Profits)。
12. 契約之期限 (Term of Agreement)。
13. 契約之終止 (Termination of Agreement)。
14. 糾紛之仲裁 (Arbitration of Disputes)。
15. 經理人之更換 (Substitution of Manager)。
16. 帳簿記載 (Books of Account)。
17. 權益轉讓之限制 (Limitations on Assignment of Interest)。
18. 認購之撤回 (Withdraw of Subscribers)。
19. 營業之清算 (Liquidation of Business)。
20. 準據法 (Governing Law)。

CONTRACT

1. Obj... ...his Contract

1.1. ...tomer shall order and the Executor shall
...consu... ...der the Technical Assignment (Ap...
Contra... ...ages) of the performance of
1.2. Per... ...nment.
Technica...

...e Parties

2. Obligatio... ...shall be obliged:
2.1. The Cu... ...he work perform...
a) to pa... ...mely all nece...
Contract... ...
b) to pro... ...o provide f...
Executor, ...be oblige...
c) if necess... ...under th...
2.2. The Executo... ...e of wor...
a) to perform... ...the ful...
b) upon perfo... ...ntract a...
c) to perform v... ...ontract a...
determined in th...

3. Procedure of Work
3.1. The Executor shall ...m work ...
...however, subject to terms ...conditions o... ...any third ...
3.2. The Executor may e... ...pproval or not...
...led for in this Contract ...
...not required. ...tion of work and within the period 1...
...the Report on the performed work an...
...the Customer for signing. ...of receipt of the Report an...
...y days of the date of receipt of the Report shall be dec...
...ion, the performed work shall be...
...shall be argumented and inclu...
...the assignment. In such...

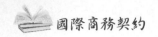

合資契約

本協議係於二〇〇〇年二月二十五日，由下列各造所訂立：

A 公司，係依中華民國法律設立，主營業所位於：＿＿＿＿＿

B 公司，係依英國法律設立，主營業所位於：＿＿＿＿＿

前言：

一、緣 A 公司（如後述之定義）計劃依據中華民國法律設立「人壽保險股份有限公司」，以經營人身保險事業（下稱「人壽業」）。

二、B 公司（如後述之定義）已同意參與投資本公司之設立，並願與 A 公司合作，以移轉有關「人壽業」發展及營運之專業技術及經驗。

基於此等原因及本約所含之相互協議及承諾，協議訂定人茲同意下列各點：

第一條　公司名稱

本公司名稱定為「＿＿＿＿人壽股份有限公司」。

第二條　公司章程

1.本公司應受中華民國法律、「公司章程」及本協議規定所規範。「公司章程」應以中文書寫另附簽證英文翻譯，並應置於公司之主營業所。本協議與「公司章程」之間若互有牴觸時，在不

JOINT VENTURE AGREEMENT

THIS ARGEEMENT is made as of February 25, 2000, by and among:
A CORPORATION, a corporation organized under the laws of the Republic of China having its principal place of business at _____, B CORPORATION, organized under the laws of England having its principal place of business at

_____.

WITNESSETH:

A.WHEREAS the A Corporation plan to incorporate a company limited by shares (the "Corporation") under the laws of the Republic of China to be called _____ Life Insurance Corporation to engage in the business of life insurance enterprise (the "LI Business") as provided and defined in the Insurance Act.

B.The B Corporation (as hereinafter defined) have agreed to invest and participate in the incorporation of the Corporation and to cooperate with the A Corporation transfer certain technical know-how and expertise to the Corporation in the field of development, management, and operation of the LI Business.

NOW THEREFORE, in consideration of the premises and of the mutual agreements, covenants and arrangements herein contained, the Parties hereby agree as follows:

1.Name of the Corporation

   The name of the Corporation shall be _____ Life Insurance Corporation.

2.Articles of Incorporation of the Corporation

   (A) The Corporation shall be governed by the laws of the R.O.C., the Articles of Incorporation and the Provisions of this Agreement. The Articles of Incorporation shall be written in Chinese with a certified English

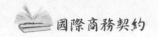
違背公司法之程度內，且就「協議約定人」之間及其本公司間
而論，應依本協議為主。中文版之「公司章程」與英文版之「公
司章程」若互有牴觸，則應以中文版為準。

2. 「協議訂定人」同意，於收到如本協議所稱之財政部之許可時，
促成本「公司」儘速設立；而「股東們」於「公司」設立後應
立即促使「公司」為下列之事項：

(1)通過「公司章程」以取代所有先前之章程。(2)決議認可本協
議，並同意受本協議拘束及行使本協議所賦予之權利。

第三條　宗旨
本「公司」之設立係為經營「人壽業」，並為經營此種事業，而具
備應有權為必要相關之行為之權，此外，本「公司」得經「董事
會」依本協議之規定決議通過從事中華民國法律所允許之其他與
「人壽業」有關之事業。

第四條　資本
本「公司」之登記資本額應為新臺幣七億元，分為記名之普通股
七千萬股，每股面額為新臺幣十元。「股份」於設立時應以現金全
數認足，「股票」以已冊錄形式印製及依「法案」規定之方式交
付。

translation, and shall be maintained at the principal office of the Corporation. In the event of there being any conflict between the provisions of this Agreement and the Articles of Incorporation then, as between the Parties and the Corporation, the provisions of this Agreement shall prevail. In the event of there being any conflict between the Chinese and the English versions of the Articles of Incorporation, the Chinese version shall prevail.

(B) The Parties agree to cause the Corporation to be formed as soon as possible following the receipt of MOF's approval as described in this Agreement and the Shareholders shall, forthwith after incorporation of the Corporation, cause the Corporation to

(a) adopt the Articles of Incorporation to the exclusion of all previous articles of incorporation of the Corporation and (b) adopt and ratify this Agreement and agree to be bound by, and entitled to exercise the rights granted to it under, its provisions.

3. Purpose

The Corporation shall be formed for the purpose of carrying on the business of a life insurance enterprise and shall have all powers necessarily incidental to the carrying on of such a business. The Corporation may, in addition, engage in such other businesses as the Directors may, subject to and in accordance with this Agreement, resolve and be conveniently carried on with its then existing business, provided the same is authorized under the laws of the R.O.C.

4. Capitalization

The total authorized capital stock of the Corporation shall be seven hundred million New Taiwan Dollars (NT$700,000,000), consisting of 70,000,000 shares of common stock, all of one series, each with a par value of NT$10.00. All Shares shall be issued upon incorporation of the Corporation,

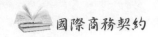

第五條　股份之認購
　　　1.於「認股完成日」，「協議訂定人」應已依其所認之「股份」，以
　　　　新臺幣現金繳足股款由「公司」收訖。「股份」之認購應如下：

「股東」姓名：　　　　　　　　　　　　　股份持有比率

　　　2.於「認股完畢」七日內：
　　　　(1)「協議訂定人」應召開股東會，在股東會應決議之事項為：

　　　　　　①選出「本國董事們」及「外國董事們」；
　　　　　　②選出「監察人」；
　　　　　　③通過「公司章程」；
　　　　　　④發行依本條第二項之規定認購之「股票」；及

　　　　　　⑤通過迄今至「公司設立日」期間之初步預算。

　　　　(2)「協議訂定人」應促成「董事會」之召開由全體董事或其委
　　　　　託書代理人出席董事會。
第六條　政府許可
　　　1.「A公司」茲於此允諾「B公司」將盡其所能促使「公司」採
　　　　取一切必要之行動以取得設立「公司」及經營「人壽業」所需
　　　　之一切中華民國政府及主管機關之許可。「B公司」應妥善協助
　　　　「A公司」取得前述之政府許可。各「協議訂定人」彼此及與
　　　　「公司」之間應相互取得前述之政府許可。

fully paid up in cash and Certificates of the Shares shall be in registered form only and to be delivered according to the Act.

5. Subscription for Shares

(A) On the Completion Date each Party shall subscribe in its own name and pay to, which shall have been received by, the Corporation in cash in New Taiwan Dollars for the full amount of the par value of the Shares initially to be subscribed to by it as set out below:

Name of Shareholder:　　　　　　　　　　　　　　Holding Percentage

(B) Within 7 days after the Completion Date:

(ⅰ) The Parties shall procure that a meeting of Shareholders is held at which:

(a) the Domestic Directors and the Foreign Directors shall be elected;

(b) the Supervisor shall be elected;

(c) the Articles of Incorporation shall be adopted;

(d) the issue of the Shares subscribed pursuant to paragraph (B) above shall be approved; and

(e) a preliminary budget for the period to the Incorporation Date shall be approved.

(ⅱ) The Parties shall procure that a board meeting shall be held to be attended by all Directors or their respective proxies.

6. Governmental approvals

(A) A Corporation hereby undertake to B Corporation that they will, and will procure that the Corporation will, take all necessary action to obtain from the R.O.C. Government or any agency or instrumentality or authority thereof any and all necessary governmental approvals required for the incorporation of the Corporation and for the Corporation to conduct the

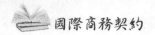

2.在不影響上項概括性之下：

⑴於本協議簽署後，「A公司」應即向「財政部」申請准許籌設「公司」進行「人壽業」。

⑵於前收到前款之許可後至遲於十日內，「外國股東們」應向經濟部投資審議委員會及其他適當之政府主管機關申請與設立「公司」有關之僑外投資許可，「A公司」並應協助「B公司」獲得此種許可。

⑶於獲得⑴款中之許可後，應連同「公司章程」向投資審議委員會申請「公司」設立執照（「公司執照」）。

⑷於獲得「公司執照」後，「A公司」須促使「公司」向「財

LI Business. B Corporation shall provide all reasonable assistance to the A Corporation and the Corporation, in obtaining from the R.O.C. Government or any agency, instrumentality or thereof any and all necessary governmental approvals required for the incorporation of the Corporation and for the Corporation to conduct the LI Business. Each Party shall cooperate reasonably with any other Party and the Corporation's efforts to obtain such approvals. Each Party shall be responsible for obtaining all governmental approvals required by it, or required by any other Party to invest in the Corporation, but which must be applied for by such first mentioned Party, as soon as possible after the necessity for such approval arises or is identified.

(B) Without prejudice to the generality of the foregoing:

(ⅰ) Forthwith on the signing of this Agreement, A Corporation shall apply to the MOF for its authorization for the establishment of the Corporation to engage in LI Business;

(ⅱ) Forthwith on the receipt of MOF approval described in this Agreement above and in any event no later than ten days thereafter, B Corporation shall file the Foreign Investment Application with the Investment Commission ("IC") of the Ministry of Economic Affairs and any other appropriate governmental authorities for all approvals required in connection with incorporation of the Corporation and Domestic Shareholders shall cooperate in providing assisstance to the Foreign Shareholders in obtaining such approvals;

(ⅲ) Forthwith on receiving the authorization applied for in subparagraph (ⅰ) above, the Parties shall cause the filing with the IC the Articles of Incorporation of the Corporation and apply for a Certificate of Incorporation of the Corporation ("Certificate of Incorporation");

(ⅳ) Forthwith on obtaining the Certificate of Incorporation, A Corporation

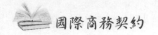

政部」申請經營「投信業」之許可（「營業執照」）。

(5) 於獲得「營業執照」後，「本協議訂定人」須促使「公司」向臺北市政府申請營利事業登記證，使公司可於符合稅法之規定下經營「人壽業」。

第七條　認股款項之使用

「協議訂定人」依本契約所繳納之股款悉應開戶存於中華民國持照之商業銀行，連同其所生之利息，僅能依下列方式運用：

1. 於核准設立前，為支付於本協議成立前經全體「協議訂定人」同意為「公司」之利益向財政部會申請核准所生之費用時，方能自前述帳戶中提領款項。

2. 於核准設立後，自集日前，僅可依「第一次股東會議」通過支出之預算自前述帳戶中提領。

第八條　「董事會」

1. 「公司」營業應由「董事會」統籌監督控制。「董事會」應享有各種依「法案」，除受本協議之不同規定外，一般應有之權利及特權。

2. 第一屆「董事」應於「第一次股東會」中選派擔任。

3. 除本協議或「公司章程」另有規定外，「公司」不得（且各「協議訂定人」均同意採取必要之行動以確保「公司」不會）為下列任一行為：

shall procure that the Corporation applies to the MOF for an operation license authorizing the Corporation to conduct LI Business ("Operation License"); and

(v) Forthwith on obtaining the Operation License, the parties shall cause the Corporation to apply for a Certificate of Profit-seeking Enterprise from the Taipei Municipal Government so that the Corporation can conduct LI Business for tax purposes.

7. Use of subscription moneys

The subscription moneys paid by the Parties in this Agreement shall be paid into an interest bearing account of the Corporation to be opened with a R.O.C. fully licensed commercial bank on terms that:

(A) prior to the issue of the Certificate of Incorporation, the only moneys that may be drawn from such account are moneys required to repay moneys expended on behalf of the Corporation in accordance with the preparatory office budget for the period to grant of the MOF authorization approved by all the Parties prior to the date of this Agreement; and

(B) after the issue of the Certificate of Incorporation, moneys may only be drawn from such account in accordance with the budget approved by the First Shareholders' Meeting.

8. the Board

(A) The operations of the Corporation shall be under the general guidance and control of the Board, which shall have, subject as varied by this Agreement, the rights and privileges normally pertaining thereto provided in the Act.

(B) The first Directors shall be appointed at the First Shareholders' Meeting.

(C) Notwithstanding any provision to the contrary in this Agreement or the Articles of Incorporation, the Corporation shall not (and each of the Parties undertakes to take all such steps as are necessary to ensure that the

⑴承擔債務或不確定責任（包括保證）或其他賠償擔保或質押，但因透支、資本認證所負擔之債務，於六個月期間總數超過「公司」資產淨值百分之十者（根據「公司」現存最新經稽核過之資產負債表並依一般會計準則計算之）；

⑵取得或處分價值超過「公司」總資產淨值百分之十之資產（依現存「公司」現存最新經稽核後之資產負債表所示），或於十二個月內，取得或處分任何公司資產，而該資產與年度已取得或已處分之資產合計超過「公司」總資產淨值百分之十者（依上述同件所示）；

⑶簽訂任何非「公司」日常營業所需之重要契約，其金額總數超過「公司」總資產淨值百分之十者（根據現存「公司」最新經稽核過之資產負債表並依一般會計準則計算之）；

⑷通過或更改任何會計政策；

⑸指定非為國際承認之會計師事務所替任為「簽證會計師」。

4.「董事」若有缺額，應由指派原董事之「協議訂定人」提名其他代表人擔任之。其他「協議訂定人」應採取必要之行動以使該新指定之代表人得順利當選為「公司」之「董事」。新任「董事」應補足原任「董事」之任期。

Corporation does not) do any of the following:

(i) incur any indebtedness or capital commitment or contingent liability (including a guarantee) or give any indemnity or security in an amount, in aggregate, over a period of twelve months, exceeding 10 percent of the net asset value of the Corporation (calculated in accordance with generally accepted accounting principles by reference to the latest audited balance sheet of the Corporation, if any);

(ii) acquire or dispose of any asset representing more than 10 percent of the net asset value of the Corporation (calculated in accordance with generally accepted accounting principles by reference to the latest audited balance sheet of the Corporation, if any) or, in any period of 12 months, any asset which, when taken together with all other assets acquired or, as the case may be, disposed of in such period, amounts to more than 10 percent of the net asset value of the Corporation (Calculated as aforesaid);

(iii) enter into any material contract, otherwise than in the ordinary course of the Corporations business involving any amount, in aggregate, exceeding 10 percent of the net asset value of the Corporation (calculated in accordance with generally accepted accounting principles by reference to the latest audited balance sheet of the Corporation, if any);

(iv) adopt or alter any accounting policy;

(v) replace the CPA by appointing anyone other than an internationally recognized firm of accountants;

(D) If for any reason any vacancy occurs in the Board, such vacancy shall be filled by an individual nominated by the Party or Parties who had nominated the person to be replaced, and all other Parties shall take such actions as may be necessary to cause the appointment or election of such

5. 「董事」得由原指派其當選之「協議訂定人」解任之。其他「協議訂定人」應依前開「協議訂定人」所請採取必要行動以解任該「董事」。

6. 除經「股東們」另有決議外，所有「董事」（包括「董事長」）及「監察人」均非專長且不領任何薪酬。「公司」無須補償「董事」因出席「董事會」所生之差旅膳宿及其他有關費用，但應支付出席費，其金額由「董事會」提議交「股東會」核准訂定之。

第九條　　「監察人」

「監察人」由「A 公司及 B 公司」提名，經「股東們」依「法案」規定過半數決議選任之。「監察人」應依「法案」執行監察權，並只有「法案」所定之職權。

第十條　　「公司」之管理

1. 「董事長」為「公司」之法定代表人，有權依「董事會」決議及本協議之規定以「公司」名義代表「公司」簽署文件，並為「董事會」及股東會會議之主席。

2. 「總經理」為公司最高執行主管，並依「董事會」之指定與監督負責「公司」日常業務行政等事宜。「總經理」應將業務行政上重要之新發展報告「董事長」及「董事會」，並應具備「營業計劃」提交「董事會」核可。「總經理」應於每次「董事會」出

nominee to the Board. The new nominee shall serve the Balance of the term of his predecessor.

(E) A Director may only be removed by the Party or Parties that nominated him and each of the other Parties shall, if requested by such Party or Parties, take such actions, as may be necessary, to cause the removal of such Director.

(F) Unless otherwise determined by the Shareholders, all Directors (including the Chairman) and Supervisor(s) shall serve on a part-time basis and without compensation. The Corporation shall not reimburse the Directors for out-of-pocket travelling, lodging, food, and incidental expenses incurred in connection with attendance at meeting of the Board but shall pay attendance fees to Directors in an amount determined, upon recommendation from the Board, by the approval of the Shareholders.

9. The supervisor

The supervisor shall be nominated by the A & B Corporation and be elected by resolution of a majority of the Shareholders in accordance with the Act. The supervisor shall perform the functions required by, and described in, the Act but shall have only those powers described in the Act.

10. Management of the Corporation

(A) The Chairman shall be the official representative of the Corporation. He shall have the right to execute documents in accordance with the resolutions of the Board and otherwise in accordance with the provisions of this Agreement in the name and on behalf of the Director. He shall preside over the meetings of the Shareholders and the Board.

(B) The President shall be the chief executive officer of the Corporation and shall, subject to the direction and control of the Board, manage the day-to-day business and operations of the Corporation. The President shall keep the Chairman and the Board advised on a current basis of all

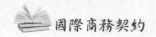

席報告營業狀況及其他對「公司」有影響之重大事項。

第十一條　「股東」

1. 「股東」常會應依「公司章程」規定召開。「股東」臨時會應由「董事會」或任何「股東」依「公司章程」之規定合法通知後召集之。

2. 「認股完成」後，「協議訂定人」全體應使「公司」不得為下列任何行為，除非業經代表已發行「股份」總數三分之二以上之「股東」同意；

   (1) 修改「公司章程」。

   (2) 辦理「公司」之增資或減資（包括借貸資本）。

   (3) 變更營業性質或範圍，或經營新營業項目。

   (4) 除「公司」破產或即將破產外，進行「公司」之解散或清算。

   (5) 出售「公司」之主要資產或合併「公司」。

第十二條　「股份」之轉讓

除轉讓予其握有股權百分之三十以上者，或新股之認購權，非

significant developments relating to the business and operations of the Corporation and shall prepare the Business Plan for approval by the Board. At each Board meeting, the President shall present to the Board review of the Corporation's operations and shall report to the Board on any material matters affecting the Corporation.

11. The Shareholders

(A) Regular Shareholders' meetings shall be called as specified in the Articles of Incorporation. Special Shareholders' meetings shall be called by the Board, or by any Shareholder, by complying with the notice and other procedures with respect thereto set forth in the Articles of Incorporation. Such notice shall contain a detailed agenda of the matters to be discussed or resolved upon at any such meeting unless a majority of the Shareholders present at such meeting agree otherwise; provided that in any event no matter may be discussed or resolved if it has not been included on the agenda received by the Shareholders.

(B) After completion the Parties hereto shall procure that the Corporation does not undertake any of the following except with an affirmative vote by Shareholders representing a two-third of the votes of the Shareholders:

(i) amend the Articles of Incorporation;

(ii) increase or reduce the capital (including loan capital) of the Corporation;

(iii) change the nature or scope of its business or commence any new business;

(iv) commence a dissolution or liquidation of the Corporation, except where the Corporation is or is likely to be, insolvent;

(v) sell substantial assets of, or merge, the Corporation.

12. Transfer of Stocks

None of the parties hereto shall sell, transfer or mortgage any of their

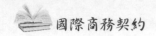

經他方同意，任一當事人不得出售或設質其股份予他人，但本條不得解釋於合併時，將股份售予繼受之公司。又受讓人須同意承受本契約及其他相關契約之責任，始得轉讓股份。

第十三條　發行新股

除中華民國之法規別有規定外，「公司」於發行新股時，「股東」有權依其持股比例優先承購。

第十四條　財務報表及會計年度

1.「公司」須：

⑴ 於會計年度開始後每季結束後 20 天內備妥每季財務報表。此報表須於備妥後儘速傳閱各「董事」，並於傳閱完畢後召開之「董事會」中納入考慮；

⑵ 會計年度終了後 60 天內備妥經「簽證會計師」查核簽證之年度財務報表，此項報告於備妥後儘速傳閱各「董事」，並於傳閱完畢後召開之「董事會」中納入考慮。

shares of the stock of the New Company without the previous written consent of the other parties except that any party may transfer its shares, or its right to subscribe to new issues, to any of its associated companies over which it has control through the direct or indirect ownership of at least thirty (30%) per cent of such associated company's voting stock. The provisions of this paragraph shall not be construed as restricting the right to any party to transfer its shares of the New Company stock to a corporation succeeding to all of its assets and obligations in connection with a merger, amalgamation or sale of substantially all of its assets. However, no such transfer shall be made unless the transferee shall agree to assume all the obligations of the transferor under this Agreement and related agreement.

13.New issues of shares

　　Subject to R.O.C. laws and regulations, any further shares in the capital of the Corporation proposed to be issued shall first be offered to the existing Shareholders pro rate to their respective holdings.

14.The financial statements and financial year

　(A) The Corporation shall prepare:

　　（ⅰ) within 20 days following the end of each three-month period following the commencement of each financial year of the Corporation, quarterly financial statements for the Corporation. Such statements shall be circulated to all the Directors as soon as practicable after the preparation thereof and shall be considered at the Board meeting next following such circulation; and

　　(ⅱ) within 60 days following the end of each financial year of the Corporation, audited annual financial statements of the Corporation. Such audited annual financial statements shall be circulated to all the Directors and supervisor as soon as practicable after the preparation thereof and shall be considered at the Board Meeting next following

2. 「公司」應備妥依「法案」、財政部或其他中華民國政府主管機關所要求及日常業務活動所必需之簿冊及紀錄。

3. 任一「協議訂定人」有權於合理通知後，於日常營業時間內檢查「公司」之簿冊及紀錄。於認為必要時，得以其自付費用對此種簿冊或紀錄為稽核或其他之必要調查（包括影印或摘要）。

4. 「公司」之帳目、簿冊及紀錄除非經「董事會」另有決議外，須一貫適用中華民國及美國，若該等美國通行之會計原則與中華民國之會計原則不相牴觸時，境內所廣泛採用之會計原則編列並保存。

5. 除第一年之會計年度自「公司」設立日起至同年之十二月卅一日止外，每一年度自每年的一月一日起至同年之十二月卅一日止。

第十五條　股息

「協議訂定人」應促使「公司」於每會計年度將當年度於彌補前一年度之虧損、繳納稅款及提撥法定盈餘公積後可分配之盈餘依下列順序及比例分派：

1. 以「股票」之實收面額不超過百分之五作為發放「股東」之股息。

2. 以可分配盈餘減去前一款之股息後計算百分之一作為員工紅利。

such circulation.

(B) The Corporation shall keep all books and record required under the Act and by the MOF and other R.O.C. governmental authorities and by good business practice.

(C) Each Party shall be entitled upon reasonable notice to examine the books and record of the Corporation during normal business hours and may, at its expense, conduct such audits and other investigations (including the taking of copies and notes) of such books and records as it shall think fit.

(D) All accounts, books and records of the Corporation shall, except to the extent the Board otherwise determines, be prepared and kept in accordance with accounting principles generally accepted in the R.O.C. and, except to the extent they are inconsistent with such R.O.C. generally accepted accounting principles, the United Stated of America, consistently applied.

(E) Each financial year of the Corporation shall commence on 1st January in each year and shall end on 31st December in that year save that the first financial year of the Corporation shall commence on the date of its incorporation and shall end on the 31st December next following the date of its incorporation.

15.Dividends

The Parties shall cause the Corporation to, after having covered losses incurred in the previous year and having provided for taxes and statutory surplus reserve(s), distribute the net profit for each fiscal year in the following priority order and proportion.

(A) Up to five percent (5%) of the paid up par value of each share as Shareholders' capital interest;

(B) One percent (1%) of the remaining distributable profit after Shareholders' capital interest set forth in sub-clause (A) above as remuneration to the

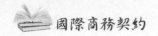

3.其餘未分配盈餘之全部或一部應由全體「股東」決定依本協議書處理。

第十六條　準據法

本協議及其每一部分應依中華民國法律解釋，並以中華民國之法律為適用之準據法。

第十七條　仲裁

「協議訂定人」之間因本協議而起或與本協議有關連之任何爭議或糾紛，如依任一「協議訂定人」之個別判斷，不能經由雙方善意的協調解決，則應由爭議之一方對他方提起且對此他方有拘束力之仲裁程序解決之。該糾紛或爭議應依國際商會調解仲裁規則於新加坡舉行之仲裁程序終局結案。仲裁人一人，應係該商會之國際仲裁法院就其獨有判斷自嫻熟財經實務者中選定。仲裁程序與仲裁判斷書應使用英文，並得於任何有管理權之法院執行。

第十八條　期間與終止

1.本協議應即日生效，並繼續有效到本協議規定期滿或「公司」解散時全部資產分配完畢兩者中先發生為止。本協議之效力不受「公司」之設立或存續，或「公司章程」之登記。

2.本協議終止時，其規定之權利義務亦隨之終止。「協議訂定

Employees; and

(C) The balance shall distributed to Shareholders subject to the conditions set forth in this Agreement.

16. Governing law

This Agreement and each part hereof shall be governed by and construed in accordance with R.O.C. law.

17. Arbitration

Any dispute among the Parties hereto arising out of or in connection with this Agreement, that cannot, in the sole judgment of any party to the controversy or dispute, be resolved through mutual negotiation in good faith, shall be resolved by arbitration proceedings instituted by one or more parties against one or more of the other Parties which shall be binding on the other Parties. Such dispute or controversy shall be finally settled by arbitration proceedings in Singapore under the Rules of Conciliation and Arbitration of the appointed by the International Chamber of Commerce by one arbitrator to be appointed by the International Court of Arbitration of the said Chamber from among persons who in the exclusive judgment of the International Court of Arbitration, are familiar with the practices of the financial business. The language of the arbitration proceedings and any award shall be in English, and may be enforced in any court having jurisdiction.

18. Term and termination

(A) This Agreement shall become effective on the date hereof, and shall continue in force until terminated pursuant to the terms hereof, or until a final distribution of all assets upon the dissolution of the Corporation, whichever occurs first, and shall not be affected by the incorporation or existence of the Corporation, or the filing of the Articles of Incorporation, or the receipt of any of the licenses, certificate or approvals.

(B) Upon termination of this Agreement, all of the rights and obligations

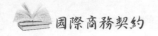

人」得以書面協議終止本協議。惟任一「協議訂定人」依本協議因他方之作為或不作為致生之損害而對該他方得有之請求權，於本協議終止後繼續生效。

第十九條　保密義務

除本協議另有規定，或中華民國或英國或其他對任一「協議訂定人」有管轄權之國家之相關法令有規定外，「協議訂定人」同意「協議訂定人」之間因相互談判而獲取之資訊或將來自本「公司」或任一「協議訂定人」所得之資訊，均應嚴格守密。除依本協議所定並係為促進「公司」之利益及目標者外，該等資訊不得為非「協議訂定人」之利益而使用，亦不得係為任一「協議訂定人」目的而使用，惟上述義務於下列情形不適用之： 1.取得該等資訊之人於取得該等資訊前即已知悉該等資訊，且係非因本協議或其相關協議衍生而得。 2.該等資訊於揭露之時已屬公共所有。3.該等資訊依本協議揭露後，非因受資訊者之過，經由出版或其他方式，成為公共所有之一部分。 4.自合法擁有該等資訊之第三人處取得，且該第三人對該等資訊最原始之提供者並無保密義務。

第二十條　轉讓

除本協議另有規定外，任一「協議訂定人」均不得將其在本協議下全部或部分之權利、利益或義務轉讓予他人。

第廿一條　協議之整體性

　1.本協議及其附件，構成協議之全部及「協議訂定人」對本協

hereunder shall terminate. The Parties hereto may by mutual consent terminate this Agreement in writing and except that all claims of any party hereto against any other party for damage arising out of acts or omissions of such other party under this Agreement shall survive such termination.

19.Confidential nature of information

As may otherwise be provided in this Agreement, the Parties agree to keep in strict confidence any information conveyed to, or acquired by, them during any negotiations between or among two or more of them, and any information obtained in the future from the Corporation or from any Party. No such information shall be used for the benefit of anyone not a Party or the purposes of any Party, except for the purpose of furthering the interests and objective of the Corporation as contemplated in this Agreement. The foregoing obligations do not, however, apply where: (1) the information was known to the person receiving it prior to the date thereof, and was not obtained or derived under this Agreement or any other agreement contemplated herein; (2) the information has already been, at the time of disclosure, in the public domain; (3) the information, which after disclosure hereunder, becomes part of the public domain, by publication otherwise, through no fault of the person receiving information; or (4) the information is obtained party in lawful possession of such information, under a confidentiality obligation to the person information originated.

20.Assignment

Except as provided in this Agreement, none of the Parties may assign, transfer or otherwise convey its rights, interest or obligations hereto, herein or hereunder in whole or in part of any other person.

21.Entire agreement

(A) This Agreement, together with the Appendices hereto, constitutes the

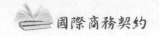

議所定一切事項上之合意。「協議訂定人」間先前之談判、聲明、陳述、承諾或合意，如有與本協議或其附件相牴觸者，均屬無效。

2.本協議內容之修正或任何協議條款之放棄，非經「協議訂定人」或其合法授權之人簽署於書面者，均屬無效。

第廿二條　使用文字

1.本協議以中、英文本簽署並交付之。如中、英文二本互相歧異時，以英文本為準。

2.所有與本協議有關之文件及交易均應以中文為之，並檢附其英文翻譯（該等翻譯之費用由「公司」負擔）。「公司」致函各「協議訂定人」時，應以英文為之。「董事會」及「股東」決議應以中文為之並檢附其英文翻譯。

第廿三條　通知

1.所有依本協議應交付予「協議訂定人」或「公司」之通知、同意、請求、合意及其他文件，應自「協議訂定人」或「公司」收受時起生效，且皆應以書面為之，以專人送達應受送達之一方之高級職員或「公司」，或以電傳、傳真或其他電訊方法為之，或，除股東會（不包括股東年會）或「董事會」之開會或其他等之延期之通知外，以航空掛號信件預付郵資方式，寄至「公司」主營業所或下列「協議訂定人」之住址：
協議訂定人：

entire agreement and understanding between the parties hereto with respect to all matters provided herein, and supersedes and cancels all previous negotiations, statements, representations, undertakings, and agreements, if any, heretofore made between or among two or more of the parties hereto.

(B) No modification or amendment of this Agreement or any of the terms or conditions hereof or thereof valid or binding, unless made in writing and signed by the Parties or a duly authorized office thereof.

22.Controlling language

(A) This Agreement has been signed and delivered in the Chinese and the English language. In case of conflict or discrepancy between the Chinese version and the English version, the English version shall always prevail.

(B) All documents prepared with respect to this Agreement and transactions contemplated herein shall be prepared in the Chinese language with English translation (the costs of such translation shall be borne by the Corporation). The Corporation, when writing to the parties hereto, shall conduct its communications in the English language. The minutes of the Board and Shareholders shall be written in the Chinese language with English translation.

23.Notice

(A) All notice, consents, requests, agreements and other documents authorized or required to be given or made by or pursuant to this Agreement to any Party or the Corporation, shall be effective upon receipt and shall be in writing, either personally served on an officer of the Party to whom it is given or the Corporation or sent by telex, facsimile or other means of electronic communication or, except with respect to a notice with regard to a Shareholders' or Board meeting or any postponement thereof (other than a regularly scheduled annual Shareholders' meeting),

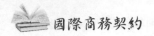

　　　或任一「協議訂定人」或「公司」隨時通知之其他住址。

　　2.前述通知及其他聯繫均應以英文為之。

　　各「協議訂定人」均已於本協議首頁日期前揭由其合法授權之代表人簽署並交付本合約，謹此聲明。

A 公司　　　　　　　　　　　　　　　B 公司
　（簽名）　　　　　　　　　　　　　　（簽名）
　（職稱）　　　　　　　　　　　　　　（職稱）

　　　　　　　　　　日 期　　　年　　月　　日

mailed by registered airmail, postage prepaid, addressed to the Corporation at its principal office or to that Party as follows: or to such other address as any such Party or any other Party or the Corporation may from time notify the other Parties and the Corporation.

(B) All such notices and other communications shall be English.

IN WITNESS WHEREOF, the parties hereto have caused their duly authorized person to have executed and delivered this Agreement as of the date first above written.

A Corporation：

_____
Signature
(Title)

Date

B Corporation：

_____
Signature
(Title)

§22 本國公司章程 (Articles of Incorporation)

按公司章程為公司之基本大法，公司之運作除公司法外，必須遵守公司章程，我國公司章程多為制式，其主要之記載內容如下：

1. 公司名稱 (Trade Name)。
2. 公司業務 (Scope of Business)。
3. 主事務所 (Principal Office)。
4. 通知 (Public Announcement)。
5. 資本額 (Authorized Capital)。
6. 股票 (Certificates of Shares)。
7. 股份之轉讓 (Transferring of Shares)。
8. 股票之遺失毀損 (Lost or Damaged)。
9. 股東會 (Shareholders Meeting)。
10. 董事會 (Board of Directors Meeting)。
11. 監察人 (Supervisors)。
12. 經理人 (Managers)。
13. 會計 (Accounting)。
14. 修正 (Amendment)。

CONTRACT

1. Obj... ...**this Contract**
1.1. T... ...tomer shall order and the Executor shall
consu... ...der the Technical Assignment (Ap-
Contra... ...ges) of the performance of
1.2. Per... ...ment.
Technica...

2. Obligatio... ...**ne Parties**
2.1. The Cu... ...r shall be obliged:
a) to pa... ...mely all nece...
Contrac...
b) to pro... ...provide f...
Executor; ...be oblige...
c) if necess... ...under th...
2.2. The Executo... ...e of wo...
a) to perform ... ...the ful...
b) upon perfo... ...
c) to perform w... ...ntract a...
determined in th... ...

3. Procedure of Work
3.1. The Executor shall ... ...m work t...
...any third ...
3.2. The Executor may er... ...conditions o...
however, subject to terms ... ...pproval or not...
...ed for in this Contract... ...
...t required. ...ction of work and within the period... ...m work an...
...y the Report on the performed work and... ...
...the Customer for signing. ...
...days of the date of receipt of the Report an...
...on. the performed work shall be d...
...shall be argumented and incl...
...ly assignment. In su...

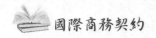

本國公司章程

第一章　總則

第一條　本公司乃依中華民國之公司法所設立之股份有限公司，本公司之
　　　　名稱為＿＿＿＿。

第二條　本公司之業務如下：

　　　　1.＿＿＿＿＿＿＿＿＿＿＿＿。
　　　　2.＿＿＿＿＿＿＿＿＿＿＿＿。
　　　　3.＿＿＿＿＿＿＿＿＿＿＿＿。
　　　　4.＿＿＿＿＿＿＿＿＿＿＿＿。
　　　　本公司業務之執行乃依相關法令辦理之。

第三條　本公司之總公司位於中華民國臺北市，得依董事會之決議，於本
　　　　國或外國設立分公司。

第四條　對本公司股東之通知必須以信函為之，如依法必須為公告者，該
　　　　通知應以總公司所在地之日報為之。

第二章　股份

第五條　本公司之資本額為新臺幣＿＿＿＿元，每股為新臺幣 10 元，所有之
　　　　股份應對外發行之。

ARTICLES OF INCORPORATION

CHAPTER I. GENERAL PROVISION

ARTICLE 1

This Corporation is incorporated as a company limited by shares under the Company Law of the Republic of China and its name is " _____ ." (registered in Chinese as " _____ ")

ARTICLE 2

The scope of business of the Corporation is as follows:

1. _____ ;
2. _____ ;
3. _____ ;
4. _____ .

The operation of the above business shall be conducted in accordance with the laws and regulations concerned.

ARTICLE 3

The Corporation shall have its head office in Taipei, Taiwan, the Republic of China, and, by resolution of its Board of Directors, may set up branches or offices at various locations within or outside the territory of the Republic of China, wherever the Corporation may deem appropriate.

ARTICLE 4

Any notification from the Corporation shall be dispatched in registered letters to all shareholders, and if public announcement is required by law. Such announcement shall be published in a conspicuous place in current newspapers of the site of the Corporation's head office.

CHAPTER II. SHARES

ARTICLE 5

The authorized capital of the Corporation shall be in the amount of _____

第六條　本公司之股票應於登記後發行之,且必須有至少 3 名董事為簽署,
　　　　並須記載公司法第一六二條各款，並依法為簽證。

第七條　股東辦理股份移轉時，必須於股票背書，並填寫股份轉讓申請書,
　　　　該股份之轉讓非登載於股東名簿前，不得對抗該公司。

第八條　如股票遺失或毀損時，股東必須立即通知該公司並於公司所在地
　　　　之當地日報為公告，亦必須於遺失或毀損地為公告。股東必須取
　　　　得除權判決，並將前述公告交與公司，公司始得補發股票。

New Taiwan Dollars (NT$ _____), divided into _____ (_____) nominative common shares, each with a par value of Ten New Taiwan Dollars (NT$10). All shares shall be issued.

ARTICLE 6

Certificates of shares to be issued after the registration of incorporation, or any new or substitute certificates of shares shall be signed and sealed by at least three directors, shall be numbered, shall contain the items specified under Article 162 of the Company Law, and shall be duly attested according to the law concerned.

ARTICLE 7

Any shareholder wishing to transfer his shares to others shall endorse his share certificate together with the transferee and fill out an application for transfer and file the same with the Corporation for entry in the roster of shareholders to complete the share transfer procedure. The transfer of shares shall not be valid against the Corporation before the completion of such share transfer procedure.

ARTICLE 8

In case a share certificate is lost or damaged, the shareholder of such certificate shall immediately report the loss or damage to the Corporation in writing and shall announce the cause of such loss or damage by public notice for three days in a conspicuous place in current newspapers of the site of the Corporation, as well as the place where such certificate of shares was lost or damaged. The shareholder shall obtain a Judgment of Exclusion confirming the loss or damage. A substitute certificate may be issued after the shareholder has complied with the above formalities and submitted a set of the newspapers containing said notice and the Judgment of Exclusion, together with an application for a substitute certificate, affixed with its original seal.

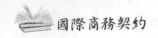

第九條　股票申請遺失或轉讓發行新股時，公司得收取一定之費用。

第十條　公司之股東必須於公司留有印鑑卡，以供公司確認股東身分或行使股東權、發放股利或其他利益之用。

第十一條　公司股份之轉讓於股東會召開前一個月停止轉讓，於臨時股東會十五天前、或於股利發放前五天前停止股份轉讓之過戶。

第三章　股東會
第十二條　股東會分為股東常會及股東臨時會。

股東常會應於每一會計年度後 6 個月內召集之。

股東臨時會得由持有公司股份超過 3% 之股東，以書面請求董事會召集之，股東且須持有公司股份一年以上。

監察人得依法召開股東會。

持有公司股份至少 3% 之股東，得經地方主管機關之同意，於

ARTICLE 9

For issuance of new share certificates due to transfer or loss of certificates, the Corporation may charge an appropriate fee for each new certificate to cover the cost so incurred.

ARTICLE 10

Each shareholder shall submit his signature or seal specimen to the Corporation for the purpose of the Corporation's verification of documents and notices issued by shareholders in exercising shareholders' rights, and receiving dividends, bonuses or other benefits.

ARTICLE 11

All entries for transfer of shares shall be suspended one month prior to each regular shareholders' meeting, fifteen days prior to each special shareholders' meeting, and five days prior to the distribution date for distributing dividends, bonuses or other benefits.

CHAPTER III.   SHAREHOLDERS' MEETING

ARTICLE 12

Shareholders' meetings of the Corporation shall be of two kinds, namely, general meetings and special meetings.

General shareholders' meetings shall be convened by the board of directors within six months after the close of each fiscal year.

Special shareholders' meetings shall be convened by a resolution of the board of directors or in accordance with written requests from shareholders representing more than three percent of the total amount of then issued and outstanding shares, provided that such shares have been continuously held by the same shareholders for more than one year.

The supervisor may convene a shareholders' meeting in accordance with the law whenever deemed necessary.

When the board of directors or the supervisor is unable to convene the

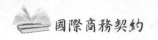

　　　　　董事會或監察人無法召開股東會時召集之。

第十三條　股東常會必須於召集前 30 日通知股東，股東臨時會須於開會前
　　　　　10 日以書面通知股東，該通知必須載明召集之事由。

第十四條　股東於公司新發行之股份以面額優先認購之，如股東放棄其優
　　　　　先認購權者，其餘股東得依其所持有比例認購該股份。

第十五條　股東就其所有之股份有其表決權，但如超過 3% 者，得以章程
　　　　　限制之，如應依比例計算而不足一股者，得以一股計算之。

第十六條　除公司法有規定者外，股東會乃以參加或經委託參加之股東多
　　　　　數為決議。

shareholders' meeting by reason of share transfers or any other causes, a shareholder who holds at least three percent of the total amount of then issued and outstanding shares of the corporation may convene the meeting with prior approval of the local competent authority.

ARTICLE 13

To convene a regular shareholders' meeting, a written notice shall be given to each shareholder thirty days in advance. To convene a special shareholders' meeting, a written notice shall be given to each shareholder ten days in advance. The written notice shall state the purposes for which the meeting is convened.

ARTICLE 14

Each shareholder shall have a preemptive right to purchase, at par value, newly issued shares of the Corporation by reason of any increase of its authorized capital or paid-in capital, in proportion to the percentage interest of each shareholder in the then issued and outstanding shares of the corporation. In the event that any shareholder waives his option or preemptive right, other shareholders shall be entitled to purchase such shares in the ratio of their participation in the Corporation.

ARTICLE 15

Each shareholder shall be entitled to one vote for each share held by him, provided that if his shares exceed three percent of the total number of issued and outstanding shares, the voting rights of the shares in excess of three percent shall be discounted by one percent. A fractional vote resulting from such calculation shall be regarded as one full vote.

ARTICLE 16

Unless otherwise provided in the Company Law, a quorum for a shareholders' meeting shall be the presence of shareholders or their duly authorized proxies representing a majority of the total amount of issued and

第十七條　公司之股東會得於本國或外國召集之。

第十八條　如股東未能參加股東會者，得以委託書委託他人參加股東會，如受託人為信託業者，得委託 2 人以上之代理人參加股東會，但其所代理出席之表決權不得超過 3%。

第十九條　董事長為股東會之主席，如董事長未能主持該股東會者，得由副董事長代理之，如公司未有副董事長或副董事長未能代理者，董事長得決定任一董事為代理人主持董事會。

第二十條　股東會之議事錄必須由董事長簽署之，且必須與股東簽名之名冊及其委託書共同存放之，股東會之記錄須於股東會結束後 15 日內寄送於股東。

outstanding shares of the corporation. All resolutions shall be made at the shareholders' meeting upon the vote of shareholders representing a simple majority of shares represented at the meeting.

ARTICLE 17

The shareholders' meeting may be held within or outside the territory of the Republic of China.

ARTICLE 18

A shareholder who is unable to be present at a shareholders' meeting may authorize a proxy by a power of attorney to represent him at the meeting, provided, however, with the exception of firms engaged in trust business, in the event that a proxy acts for two or more shareholders, his delegated voting rights shall not exceed three percent of the total issued and outstanding shares.

ARTICLE 19

The chairman of the board of directors shall preside at shareholders' meetings. In case of the absence of the chairman of the board of directors, the vice chairman of the board of directors shall preside in lieu of him, or if there is no vice chairman of the board of directors or the vice chairman of the board of directors is unable to perform his duty, the chairman of the board of directors shall designate a director to preside in lieu of him, otherwise, the directors shall elect one from among themselves to preside in lieu of the chairman of the board of directors.

ARTICLE 20

Resolutions adopted in a shareholders' meeting shall be recorded in the minutes of the proceedings which shall be signed or sealed by the chairman of the meeting. The minutes shall be kept together with the attendance book signed by the shareholders present at the meeting and the proxies. A copy of the minutes shall be distributed to each shareholder within fifteen days after the meeting.

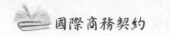

第四章　董事會

第二十一條　公司應由股東或法人之代表中選任董事三人，董事之任期為
　　　　　　3年，且得連選而連任。

第二十二條　法人股東得選任代表人為公司之董事，且法人得指派新代表
　　　　　　人為更換或承繼。

第二十三條　董事中選任其中一人為董事長，並應依此方法選任副董事長，
　　　　　　董事長為董事會及股東會之主席，並依公司法對外代表公司
　　　　　　並行使其職權，如董事長未能執行其職務時，由副董事長代
　　　　　　理之，如公司無副董事長者，董事長得指定董事中之一人代
　　　　　　理之，如未能指定者，應由董事選任其中一人代理董事長。

第二十四條　董事會應由董事長召集之，第一次董事會應由得票最高之董
　　　　　　事召集之，第一次董事會應於選任新董事即日起15日內召開
　　　　　　之，或前任董事屆期滿15日內，以最先到期者為準。董事會
　　　　　　得於中華民國國內召集之，如董事不能參加董事會者，得以
　　　　　　委託書指定其他董事代理之，任一董事僅得代理一董事。

CHAPTER IV.　BOARD OF DIRECTORS' MEETING

ARTICLE 21

The Corporation shall have three directors to be elected from among the shareholders or representatives with legal capacity designated by the corporate shareholders. The directors shall serve for a term of three years and may continue to serve if reelected.

ARTICLE 22

The corporate shareholders of the Corporation shall have the right to designate representatives to be elected as directors of the Corporation and the right to designate representatives as substitutes or successors of such directors.

ARTICLE 23

The Corporation shall have a chairman of the board of directors to be elected by and from among the directors. The Corporation may have a vice chairman of the board of directors to be elected by the same method. The chairman shall preside at the shareholders' meetings and the board of directors' meetings, and externally represent the Corporation and exercise other functions under the Company Law. In case the chairman of the board of directors is unable to exercise his functions, the vice chairman of the board shall act on his behalf, or if there is no vice chairman of the board or the vice chairman of the board is unable to exercise his functions, the chairman of the board shall designate a director to act on his behalf. Otherwise, the directors shall elect one from among themselves to act on the chairman's behalf.

ARTICLE 24

A board of directors' meeting shall be convened by the chairman of the board, provided that the initial meeting of each term of the board of directors shall be convened by the director who was elected with the greatest number of votes. The initial meeting of each new term shall be held within fifteen days after the election of the new directors or the expiry of the term of the former

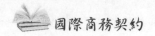

如住所於外國之董事未能參加董事會者，得以書面指派於中華民國有住所之股東定期參加董事會。

董事會之通知應記明其召集目的及議程，且應於召開之日前7日通知之，但如有緊急情事時，得隨時召集之。

第二十五條　除公司法第一八五條、二〇八條第一項、二四六條、二六六條及二八二條之規定，須以三分之二以上董事同意者外，董事會之決議以多數決之方式為之。

本章程第二十條有關股東會議事錄之規定於董事會準用之。

第二十六條　董事會之任務如下：

　1.核准公司之規則；
　2.準備公司之各項營業計劃；
　3.決議公司之會計帳冊與預算；
　4.委任或解任公司之總經理、副總經理及經理；

directors, whichever is later. Board of Directors' meetings may be held within or outside the territory of the Republic of China. Any director who is unable to attend a meeting may appoint another director to act as his proxy by a power of attorney. A director may act as proxy for only one absent director.

A director residing in a foreign country may appoint in writing a shareholder domiciled within the territory of the Republic of China to regularly attend by proxy meetings of the board of directors.

Notice of a board of directors' meeting shall specify the purpose and agenda for which the meeting is convened and shall be given seven days prior to the meeting, provided, however, that where an convened at any time.

ARTICLE 25

A majority of the directors shall constitute a quorum of the board of directors' meeting after notice of the meeting has been served on all directors, except for matters specified in Articles 185, 208 I, 246, 266 and 282 of the Company Law and for the election of the chairman and vice chairman of the board of directors, which shall require a quorum of two-thirds of the directors. Any resolution adopted by the board of directors shall require at least a majority vote of the directors present in favor of such resolution.

The provisions of Article 20 above on the minutes of the shareholders' meetings shall apply, mutates mutandis, to the minutes of the board of directors' meetings.

ARTICLE 26

The authorities and duties of the board of directors are as follows:

1. To approve import rules and by-laws of the Corporation;
2. To prepare the business plan;
3. To approve the budget and final accounting of the books;
4. To appoint and discharge the general manager, vice general managers and managers;

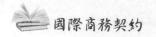

5.決議有關公司之利益與損失之分配；

6.決議有關公司之增資或減資；

7.告知公司之監察人有關公司現有或潛在之損失；

8.依公司法或股東會之決議行使其職權。

第二十七條　董事會應指派秘書處理有關依董事會之指示必須處理之公司文件。

第五章　監察人

第二十八條　公司應由股東或法人股東之代表選任一監察人，監察人之任期為 3 年，且得連選為連任。

第二十九條　法人股東得指派代表人當選為監察人，且法人得隨時指派新代表人為更換或承繼。

第三十條　監察人之任務如後：

1.檢查董事會所準備之財務報表，

2.審計及檢查公司之預算及年度財務報表，

3.檢查公司之營業，

4.檢查公司之財務帳冊、費用、收入及資產，

5. To prepare a proposal for appropriation of profits or losses;

6. To prepare a proposal for capital increase or decrease;

7. To inform the supervisor of any danger of sustaining material losses or damages to the Corporation; and

8. To exercise other powers and duties in accordance with the Company Law or resolutions of the shareholders' meetings.

ARTICLE 27

The board of directors may appoint a secretary who shall handle the important documentation for the Corporation under the direction of the board of directors.

CHAPTER V.　SUPERVISORS

ARTICLE 28

The Corporation shall have one supervisor to be elected from among the shareholders or representatives designated by the corporate shareholders with legal capacity. The supervisor shall serve for a term of three years and may continue to serve if re-elected.

ARTICLE 29

The corporate shareholders of the Corporation shall have the right to designate representatives to be elected as supervisor of the corporation and the right to designate representatives as substitutes or successors of such supervisor.

ARTICLE 30

The authorities and duties of the supervisor are as follows:

1. To audit the accounting statements prepared by the board of directors and presented by to the shareholders' meeting;

2. To audit the budget and the finalization of the accounting books;

3. To inspect the business;

4. To examine the accounting books, disbursements, revenue and assets;

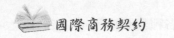

5. 通知公司董事會停止其違反法令或章程之行為，或其違反公司營業範圍之行為，

6. 依公司法行使其職權及履行其義務。

第三十一條　　監察人就其檢查之會計帳冊應為簽名，且於檢查後報告於股東會。

監察人得代表公司聘請律師或會計師就前項事務為檢查，監察人得以觀察人之身份參加董事會並為陳述，但不得為投票。

第六章　　經理人

第三十二條　　公司應指派一名總經理、副總經理、數名經理人，且得依董事會之指示僱用其他職員。

第三十三條　　公司之總經理應由董事長所提名並由董事會委任或解任，公司之副總經理及經理人應由總經理提名，並由董事會為聘任或解任。

第七章　　會計

第三十四條　　公司應以每年 1 月 1 日起及 12 月 31 日止為會計年度，編制財務報表，董事會必須就下列文件為準備，並於股東會召開前 30 日交予監察人檢查，監察人應於其檢查後或經會計師審

5. To notify the board of directors to cease doing any act performed in contravention of applicable law or regulation, or the articles of incorporation, or to cease operating any business beyond the Corporation's registered scope of business; and

6. To exercise his authority and perform other duties in accordance with the Company law.

ARTICLE 31

The supervisor shall sign or seal his chop on the accounting statements which he has audited, and shall report to the shareholders' meeting after such audit.

The supervisor may, on behalf of the Corporation, retain attorneys-at-law or certified public accountants to under take the audit referred to in the preceding paragraph. The supervisor may be present at the board of directors' meeting as an observer to make statements but shall not vote.

CHAPTER VI.　MANAGERS

ARTICLE 32

The corporation shall have one general manager, one vice general manager and several manager, and may employ other staff to handle the business under the direction of the board of directors.

ARTICLE 33

The general manager shall be nominated by the chairman of the board of directors and shall be appointed or discharged by the board of directors. The vice general manager and managers shall be nominated by the general manager and shall be appointed or discharged by the board of directors.

CHAPTER VII.　ACCOUNTING

ARTICLE 34

The Corporation shall finalize the accounting books at the end of each fiscal year which shall begin on January 1st and end on December 31st. The

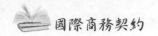

計後，於下述文件為簽署，並於接受該財務報表後，將其報告交予股東會為承認：

　1.營業報告
　2.損益表
　3.資產負債表
　4.資產目錄
　5.公司損益之分配之建議

第三十五條　於公司年度財務報表中顯示公司有盈餘者，公司先應負擔稅捐，並就其盈餘之 10% 為法定盈餘公積，並得經股東會決議為分配。

如法定盈餘公積已達其資本公積時，公司得停止提存法定盈餘公積。

第三十六條　公司得以公司盈餘之百分之＿＿＿＿為員工紅利。本條所稱公司之盈餘為應分配予股東股利之可分配盈餘部份。

公司之總經理得依其職權支付紅利予員工，但必須於年度終了後依據年度財務報表計算之，該紅利必須依法且得依稅法得認定為費用，始得分配之，如該紅利超過其淨利或股東會所承認之利益時，該超過之部份不得視為員工之特別紅利，應以公司之費用計算之。

board of directors shall prepare the following statements and refer them to the supervisor for audit at least thirty days prior to the regular shareholders' meeting. The supervisor shall seal his chop on the following statements after having audited or having deputized a certified public accountant to audit them, and shall submit a report to the general shareholders' meeting for acceptance.

1. Operations report

2. Balance sheet

3. Profit and loss statement

4. Property inventory

5. Proposal for distribution of profits or covering of losses

ARTICLE 35

In the event that the Corporation generates profits as indicated by the final accounting of the books, the Corporation shall first pay all taxes and duties, then set aside one tenth of its net profits as legal reserve before it may distribute dividends and bonus by resolution of the shareholders' meeting.

The corporation may cease to set aside its profit as legal reserve as provided in the preceding paragraph, if such legal reserve has been accumulated to equal the total paid-in capital.

ARTICLE 36

The employees' share of bonus shall be _____ % of the corporation net profit. Net profit as referred to in this article shall mean the distributive earnings of the year net of the dividends (capital interest) to the shareholders, which shall be calculated at the rate of ten percent of the paid-in capital.

The general manager of the Corporation is authorized to pay the bonus to the employees, as specified in the preceding paragraph, after the final accounting of the books of the previous fiscal year, based upon the estimated net profit. Such payment to the employees shall be regarded as an expense item for the year paid, for the corporation's business income tax purposes to the

第八章　附則

第三十七條　公司營業之相關規則得由董事會另行規定之。

第三十八條　本章程未規定者，依公司法及其法令辦理。

第三十九條　本章程經＿＿＿＿年＿＿＿月＿＿＿日之發起人會議決議之，且於主管機關同意後生效，本章程之修正應經股東會之決議，且應於主管機關為登記。

extent permitted by law. In the event the bonus paid to the employees is in excess of the amount based upon the audited net profit as approved by the shareholders' meeting, the excess portion shall not be refunded or reserved as bonus for subsequent years, but shall be treated as special bonus to the employees and shall be regarded as an expense item for the Corporation's business income tax purposes.

CHAPTER VIII.　SUPPLEMENTARY PROVISIONS

ARTICLE 37

The organizational rules of the Corporation may be separately prescribed by the board of directors.

ARTICLE 38

With regard to all matters not provided for in these Articles of Incorporation, the Company Law and other laws and regulations concerned shall govern.

ARTICLE 39

These Articles of Incorporation were adopted by resolution of the meeting of incorporation on the _____ day of _____, 20 _____, and shall become effective upon approval by the competent authorities. Any amendment to these Articles of Incorporation shall be made by resolution of the shareholders' meeting and submission to the competent authority for registration.

§23 外國公司章程 (Articles of Association)

外國公司與本國公司之章程差異，在於外國公司就各項事務有較明確之規定，其通常包括之內容與本國並無差異，主要之內容包括：

1. 股票 (Shares Certificates)。
2. 股份之發行 (Issue of Shares)。
3. 股份之轉讓 (Transfer of Shares)。
4. 股份之買回 (Redeemable Shares)。
5. 不同股份 (Variation of Shares)。
6. 公司章程之修正 (Amendment of Memorandum of Association)。
7. 停止過戶 (Closing Register)。
8. 股東常會 (General Meeting)。
9. 表決權 (Votes of Members)。
10. 董事 (Directors)。
11. 董事之代理 (Alternate Directors)。
12. 董事之權責 (Powers and Duties of Directors)。
13. 經理人 (Management)。
14. 股利 (Dividends)。
15. 審計 (Audit)。
16. 清算 (Winding Up)。

CONTRACT

1. Obj... ...his Contract
1.1. The C... ...tomer shall order and the Executor shall...
consu... ...der the Technical Assignment (Ap...
Contra... ...ages) of the performance of...
1.2. Per... ...nment.
Technica...

2. Obligatio... ...e Parties
2.1. The Cu... ...r shall be obliged:
a) to pa... ...mely all nec...
Contrac...
b) to pro... ...provide f...
Executor; ...be oblige...
c) if necess... ...under th...
2.2. The Execut... ...e of wo...
a) to perform... ...n the ful...
b) upon perfo... ...ntract a...
c) to perform... ...ntract a...
determined in th...

3. Procedure of Work
3.1. The Executor shall... ...m work...
...any third p...
3.2. The Executor may e... ...conditions o...
however, subject to terms... ...pproval or not...
...ed for in this Contract...
...required. ...ion of work and within the period in...
...Report on the performed work and...
...the Customer for signing.
...days of the date of receipt of the Report an...
...tion, the performed work shall be d...
...hall be argumented and incl...
...he performed work shall be...
...e assignment. In suc...

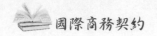

外國公司章程

第一條　股票

　　　　公司股票得由董事決定其型式，且得為公司用印。

第二條　股份之發行

　　　　公司得依公司章程、股東會之指示，且不損害原股東權益之前提下，發行股份、選擇權予他人。

第三條　股份之轉讓

　　　　本公司股份之轉讓應以書面為之，且必須由讓與人與受讓人或其代理人為之，於股東名簿變更前，原名義人仍被視為股東。

第四條　可收回之股份

　　　　依公司章程之規定，該股份得依發行條件、股東或公司之選擇，而預定而為股份之收回，該公司為收回時，應由股東會特別決議。

第五條　不同股份之權利

　　　　如公司之股份分為不同種類時，除發行條件另有規定，各種類發行之股份之權利，得依該種類股東四分之三之決議，或股東常會之特別決議而變更其權利之內容。

ARTICLES OF ASSOCIATION

ARTICLE 1.　CERTIFICATES FOR SHARES

Certificates representing shares of the Company shall be in such form as shall be determined by the Directors. Such certificates may be under Seal.

ARTICLE 2.　ISSUE OF SHARES

Subject to the provisions, if any, in that behalf in the Memorandum of Association and to any direction that may be given by the Company in general meeting and without prejudice to any special rights previously conferred on the holders of existing shares, the Directors may allot, issue, grant options over or otherwise dispose of shares of the Company.

ARTICLE 3.　TRANSFER OF SHARES

The instrument of transfer of any share shall be in writing and shall be executed by or on behalf of the transferor and the transferor shall be deemed to remain the holder of a share until the name of the transferee is entered in the register in respect thereof.

ARTICLE 4.　REDEEMABLE SHARES

Subject to the provisions of the Statute and the Memorandum of Association, shares may be issued on the terms that they are, or at the option of the Company or the holder are, to be redeemed on such terms and in such manner as the Company, before the issue of the shares, may by Special Resolution determine.

ARTICLE 5.　VARIATION OF RIGHTS OF SHARES

If at any time the share capital of the Company is divided into different classes of shares, the rights attached to any class (unless otherwise provided by the terms of issue of the shares of that class) may, whether or not the Company is being wound up, be varied with the consent in writing of the holders of three-fourths of the issued shares of that class, or with the sanction of a Special

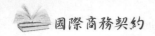

第六條　信託之不承認
　　　　除股東之全部或既有權利，公司不承認任何信託之任何權利，公
　　　　司無義務接受任何有關股權、部分股權、附加股權之股東權益。

第七條　通知股東
　　　　董事會得就股份未支付之金額，不論該金額是股票之表面價值或
　　　　其溢價，均得通知股東支付，且不受固定期間應支付期限之規定。

第八條　股份之沒收
　　　　如股東未支付或未能部份支付於發行日且應支付之資金，董事會
　　　　得隨時通知股東支付分期之款項或未支付之價金，並得請求利息。

第九條　授權文件之登記
　　　　公司得就有關認證、管理、死亡或結婚之證明、委任書或其他文
　　　　件收取不超過一美金之費用。

第十條　股份之承受
　　　　如股東死亡時，繼承人或繼承部分股份之繼承人，或股東之法定

Resolution passed at a general meeting, of the holders of the shares of that class.

ARTICLE 6.　NON-RECOGNITION OF TRUSTS

No person shall be recognized by the Company as holding any share upon any trust and the Company shall not be bound by or be compelled in any way to recognize (even when having notice thereof) any equitable, contingent, future, or partial interest in any share, or any interest in any fractional part of a share, or (except only as is otherwise provided by these Articles or the Statute) any other rights in respect of any share except an absolute right to the entirety thereof in the registered holder.

ARTICLE 7.　CALL ON SHARES

The Directors may from time to time make calls upon the Members in respect of any monies unpaid on their shares (whether on account of the nominal value of the shares or by way of premium or otherwise) and not by the conditions of allotment thereof made payable at fixed terms.

ARTICLE 8.　FORFEITURE OF SHARES

If a member fails to pay any call or installment of a call or to make any payment required by the terms of issue on the day appointed for payment thereof, the Directors may, at any time thereafter during such time as any part of the call, installment or payment remains unpaid, give notice requiring payment of so much of the call, installment or payment as is unpaid, together with any interest.

ARTICLE 9.　REGISTRATION OF EMPOWERING INSTRUMENTS

The Company shall be entitled to charge a fee not exceeding one dollar (US$1.00) on the registration of every probate, letters of administration, certificate of death or marriage, power of attorney.

ARTICLE 10.　TRANSMISSION OF SHARES

In case of the death of a Member, the survivor or survivors where the

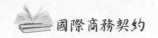

代理人，其將為唯一公司所承認享有該股份之人。

第十一條　公司章程之修正、事務所地址之變更及資本之變更
　　　　　公司章程之修正、事務所地址之變更及資本之變更依法律之規
　　　　　定，股東會得依股東之決議變更股東會章程。

第十二條　股東名簿變更之停止日期
　　　　　為決定股東於股東會表決之權限，或為分配股利於股東，董事
　　　　　會得以四十日之期限內並為停止過戶之日期。

第十三條　股東常會
　　　　　股東會應於公司成立後一年內或於每年召開股東會，而其通知
　　　　　應署名召集之事由。

第十四條　股東常會之通知
　　　　　股東常會之通知應於開會前五日為之，該通知應署名開會日期、
　　　　　時間及地點。

第十五條　股東之投票
　　　　　除特定種類股份股票權之限制者外，每一股東應出示其股份之

deceased was a joint holder, and the legal personal representatives of the deceased where he was a sole holder, shall be the only persons recognized by the Company as having any title to his interest in the shares.

ARTICLE 11.　AMENDMENT OF MEMORANDUM OF ASSOCIATION, CHANGE OF LOCATION OF REGISTERED OFFICE & ALTERATION OF CAPITAL

Subject to and in so far as permitted by the provisions of the Statute, the Company may from time to time by ordinary resolution alter or amend its Memorandum of Association.

ARTICLE 12.　CLOSING REGISTER OF MEMBERS OR FIXING RECORD DATE

For the purpose of determining Members entitled to notice of or to vote at any meeting of Members or any adjournment thereof, or Members entitled to receive payment of any dividend, or in order to make a determination of Members for any other proper purpose, the Directors of the Company may provide that the register of Members shall be closed for transfers for a stated period but not to exceed in any case forty days.

ARTICLE 13.　GENERAL MEETING

The Company shall within one year of its incorporation and in each year of its existence thereafter hold a general meeting as its annual general meeting and shall specify the meeting as such in the notices calling it.

ARTICLE 14.　NOTICE OF GENERAL MEETINGS

At least five days' notice shall be given of an annual general meeting or any other general meeting. Every notice shall be exclusive of the day on which it is given or deemed to be given and of the day for which it is given and shall specify the place, the day and the hour of the meeting.

ARTICLE 15.　VOTES OF MEMBERS

Subject to any rights or restrictions for the time being attached to any class

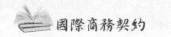

記錄或委託書，以行使其表決權。

第十六條　委託書
委託書之文件須為書面，且須由委任人於文件聲明，如委任人為公司者，必須由公司之職員或代理人為簽名，受託人不限於公司之股東。

第十七條　董事
公司得選任十名以內之董事，並得由股東會之股東決議增加或減少董事之數目，首屆董事應由股東會以書面表決，或發起人為決議。

第十八條　董事之代理
董事因故未能出席董事會者，得指定他人代理出席董事會。

第十九條　董事之權利與義務
董事得以公司之費用管理公司，並於非違反公司之章程、法規之前提下行使其權利。

or classes of shares, on a show of hands every Member of record present in person or by proxy at a general meeting shall have one vote and on a poll every Member of record present in person or by proxy shall have one vote for each share registered in his name in the register of Members.

ARTICLE 16.　PROXIES

The instrument appointing a proxy shall be in writing and shall be executed under the hand of the appointor or of his attorney duly authorized in writing, or, if the appointor is a corporation under the hand of an officer or attorney duly authorized in that behalf. A proxy need not be a Member of the Company.

ARTICLE 17.　DIRECTORS

There shall be a Board of Directors consisting of not less than one or more than ten persons (exclusive of alternate Directors) PROVIDED HOWEVER that the Company may from time to time by ordinary resolution increase or reduce the limits in the number of Directors. The first Directors of the Company shall be determined in writing by, or appointed by a resolution of, the subscribers of the Memorandum of Association or a majority of them.

ARTICLE 18.　ALTERNATE DIRECTORS

A Director who expects to be unable to attend Directors' Meetings because of absence, illness or otherwise may appoint any person to be an alternate Director to act in his stead.

ARTICLE 19.　POWERS AND DUTIES OF DIRECTORS

The business of the Company shall be managed by the Directors (or a sole Director if only one is appointed) who may pay all expenses incurred in promoting, registering and setting up the Company, and may exercise all such powers of the Company as are not, from time to time by the Statute, or by these Articles, or such regulations, being not inconsistent with the aforesaid.

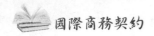

第二十條　管理
　　　　　　董事會得於其合理權限範圍內，就公司之管理提供服務。

第二十一條　常務董事
　　　　　　董事會得以公司之費用，指定一人或數人為常務董事。

第二十二條　董事會之程序
　　　　　　除章程另有規定外，董事會應決定公司業務之進行，並就其
　　　　　　董事會之開始、延長或其他之事宜訂定規則。

第二十三條　董事會董事之辭任
　　　　　　董事如以書面通知公司辭去其董事席位時，該董事職位即為
　　　　　　空缺。
第二十四條　指認或替換董事
　　　　　　公司得以股東會之決議指派或更換董事。

第二十五條　同意之推定
　　　　　　參加董事會之董事，除非其明示反對且載明於議事錄，推定
　　　　　　其同意該議案。

第二十六條　印章
　　　　　　董事會得決定由董事或董事會之委員會就公司之行為，由董

ARTICLE 20.　MANAGEMENT

The Directors may from time to time provide for the management of the affairs of the Company in such manner as they shall think fit and the provisions contained in the three next following paragraphs shall be without prejudice to the general powers conferred by this paragraph.

ARTICLE 21.　MANAGING DIRECTORS

The Directors may, from time to time, appoint one or more of their body (but not an alternate Director) to the office of Managing Director for such term and at such remuneration (whether by way of salary, or commission, or participation in profits, or partly in one way and partly in another).

ARTICLE 22.　PROCEEDINGS OF DIRECTORS

Except as otherwise provided by these Articles, the Directors shall meet together for the despatch of business, convening, adjourning and otherwise regulating their meetings as they think fit.

ARTICLE 23.　VACATION OF OFFICE OF DIRECTOR

The office of a Director shall be vacated if he gives notice in writing to the Company that he resigns the office of Director.

ARTICLE 24.　APPOINTMENT AND REMOVAL OF DIRECTORS

The Company may by ordinary resolution appoint any person to be a Director and may in like manner remove any Director and may in like manner appoint another person in his stead.

ARTICLE 25.　PRESUMPTION OF ASSENT

A Director of the Company who is present at a meeting of the Board of Directors at which action on any Company matter is taken shall be presumed to have assented to the action taken unless his dissent shall be entered in the Minutes of the meeting.

ARTICLE 26.　SEAL

The Company may, if the Directors so determine, have a Seal which shall,

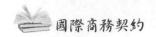

事、公司秘書、財務長等就公司之事務為用印。

第二十七條　職員

公司得經董事會之決議，指派總經理、秘書及財務長，此等職員應依其義務並享有報酬，並得依董事會隨時將其解任。

第二十八條　股利、分配及準備

依法董事會得分配年終或期終之股利，該分配須以公司對外發行可分配之資金為限。

第二十九條　公司之資本結構

公司得由董事會建議而經股東會之決議，授權董事會就公司為發行股份或為股利之分配。

第三十條　公司之帳冊

董事會應就公司之取得資金或費用設立帳戶。

第三十一條　審計

公司得隨時於股東常會中指派一人或數人為審計人員，並得

only be used by the authority of the Directors or of a committee of the Directors authorized by the Directors in that behalf and every instrument to which the Seal has been affixed shall be signed by one person who shall be either a Director or the Secretary or Secretary-Treasurer or some person appointed by the Directors for the purpose.

ARTICLE 27.　OFFICERS

The Company may have a President, a Secretary or Secretary-Treasurer appointed by the Directors who may also from time to time appoint such other officers as they consider necessary, all for such terms, at such remuneration and to perform such duties, and subject to such provisions as to disqualification and removal as the Directors from time to time prescribe.

ARTICLE 28.　DIVIDENDS, DISTRIBUTIONS AND RESERVE

Subject to the Statute, the Directors may from time to time declare dividends (including interim dividends) and distributions on shares of the Company outstanding and authorize payment of the same out of the funds of the Company lawfully available therefore.

ARTICLE 29.　CAPITALISATION

The Company may upon the recommendation of the Directors by ordinary resolution authorize the Directors to capitalize any sum standing to the credit of any of the Company's reserve accounts (including share premium account and capital redemption reserve fund) or any sum standing to the credit of profit and loss account or otherwise available for distribution.

ARTICLE 30.　BOOKS OF ACCOUNT

The Directors shall cause proper books of account to be kept with respect to all sums of money received and expended by the Company and the matters in respect of which the receipt or expenditure takes place.

ARTICLE 31.　AUDIT

The Company may at any annual general meeting appoint an Auditor or

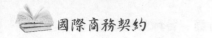

以股東會（於股東會間）之期間為審計並收取費用。

第三十二條　通知

公司對股東之通知應以書面為之，但以傳真等為之者，應補予書面文件。

第三十三條　清算

公司得經股東會特別決議或因法律之規定，而進行解散之清算。

第三十四條　賠償

公司之董事及職員因處理公司事務所受損害，應由公司資產予以補償。

第三十五條　會計年度

公司之會計年度為 1 月 1 日起至 12 月 31 日止。

第三十六條　公司章程之修正

公司得依法於股東會之特別決議修正章程。

第三十七條　讓與之延續

除法律豁免規定者外，公司得依法及股東會之特別決議於他國或屬地登記，而於本國撤銷登記。

Auditors of the Company who shall hold office until the next annual general meeting and may fix his or their remuneration.

ARTICLE 32.　NOTICES

Notices shall be in writing and may be given by the Company to any Member either personally or by sending it by post, cable, telex or telecopy to him or to his address as shown in the register of Member, such notice, if mailed, to be forwarded airmail.

ARTICLE 33.　WINDING UP

If the Company shall be wound up the liquidator may, with the sanction of a Special Resolution of the Company and any other sanction required by the Statute.

ARTICLE 34.　INDEMNITY

The Directors and officers for the time being of the Company and any trustee for the time being acting in relation to any of the affairs of the Company and their heirs, executors, administrators and personal representatives respectively shall be indemnified out of the assets of the Company from and against all actions.

ARTICLE 35.　FINANCIAL YEAR

Unless the Directors otherwise prescribe, the financial year of the Company shall end on 31st December in each year and, following the year of incorporation, shall begin on 1st January in each year.

ARTICLE 36.　AMENDMENTS OF ARTICLES

Subject to the Statute, the Company may at any time and from time to time by Special Resolution alter or amend these Articles in whole or in part.

ARTICLE 37.　TRANSFER BY WAY OF CONTINUATION

If the Company is exempted as defined in the Statute, it shall, subject to the provisions of the Statute and with the approval of a Special Resolution, have the power to register by way of continuation as a body corporate under the laws of any jurisdiction outside the Cayman Islands and to be deregistered in the Cayman Islands.

§24 技術協助契約 (Technological Assistance Agreement)

技術協助契約，又稱技術合作契約 (Technological Cooperative Agreement)，其非民法債編所訂之有名契約，但於實務上運用頗廣，其大多涉及商標權、專利權或專門技術 (Know-how) 之授與，該種契約所包括之事項如下：

1. 簽約日期 (Date of Agreement)。

2. 當事人名稱及地址 (Names and Addresses of Parties)。

3. 技術合作之目的及內容 (Purpose and Substance of Technology Transfer)。

4. 技術合作之性質及方式 (Nature, Exclusive or Non-exclusive, and Processes of Technology Transfer)。

5. 技術報酬金之支付 (Payment of Royalty)。

6. 智慧財產權之授權 (License of Intellectual Property Rights)。

7. 保密條款 (Maintenance of Secrecy)。

8. 侵害及救濟 (Infrigement and Remedies)。

9. 授權人之擔保 (Representations and Warranties of Licensor)。

10. 授權產品之品質 (Quality of Merchandise Produced under License)。

11. 被授權之違約 (Default by License)。

12. 通知 (Notice)。

13. 讓與及轉授權之禁止 (Assignment and Sublicense Forbidden)。

14. 期間及終止 (Term and Termination)。

15. 不可抗力 (Force Majeure)。

16. 管轄或仲裁條款 (Jurisdiction or Arbitration Clause)。

17. 準據法及語文 (Governing Law and Language)。

CONTRACT

1. Ob... this Contract
1.1. ...tomer shall order and the Executor shall
consu... ...der the Technical Assignment (Ap-
Contra... ...ges) of the performance of
1.2. Per... ...nment.
Technical

2. Obligatio... ...e Parties
2.1. The Cu... ...shall be obliged:
   a) to pa... ...mely all nece...
   Contrac...
   b) to pro... ...o provide f...
   Executor; ...be obliged...
   c) if necess... ...under th...
   2.2. The Executor... ...e of wor...
   a) to perform... ...n the ful...
   b) upon perfor... ...ntract a...
   c) to perform... ...ntract a...
   determined in th...

3. Procedure of Work
3.1. The Executor shall... ...rm work t...
3.2. The Executor may e... ...any third...
however, subject to terms... ...conditions o...
...ed for in this Contract... ...pproval or not...
...t required. ...tion of work and within the period th...
...e Report on the performed work an...
...the Customer for signing.
...s of the date of receipt of the Report an...
...on, the performed work shall be de...
...ssignment. In suc...
...performed work shall be argumented and incl...

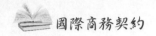
技術協助契約

本契約於一九九三年九月一日由經　　　法設置之 A 公司（主營業所於　　），其依　　　法設置之 B 公司（主營業所設置於　　），經雙方合意而訂定本契約。

前　言

授權人就所授權之物品之製造及銷售擁有若干專利、商標、技術情報及訣竅；被授權人意圖取得前述專利、商標、技術情報及訣竅之使用權，授權人亦準備授與前項之權利。因此，基於雙方對價之約束，當事人同意以下之條款：

第一條　定義

為本契約目的之達成，以下之單字及用語具有下列之意義：

⑴ 授權產品乃指（　）。

⑵ 指授權人所有有關於授權產品之專利權、使用模型及新型等權利，及於本契約期間或生效前有關政府所發給或向政府申請之技術情報。

⑶ 商標乃指授權人所有之商標或商標申請。

TECHNOLOGICAL ASSISTANCE AGREEMENT

THIS AGREEMENT, made this first day of September, 1993, by and between A COMPANY, LTD. (hereinafter called "Licensor"), a corporation organized and existing under the laws of _____, with its principal place of business at _____, and B COMPANY, LTD. (hereinafter called "Licensor"), a corporation organized and existing under the laws of _____, with its principal place of business at _____.

WITNESSETH:

WHEREAS, Licensor owns certain patents, trademarks and technical informations and know-how pertaining to the manufacture and sale of the Licensor products (as hereinafter Licensed Products), and

WHEREAS, Licensor desires to acquire the right to use aforesaid patents, trademarks and technical information and know-how and Licensor is ready to grant the same as hereinafter set forth.

NOW, THEREFORE, in consideration of the mutual pledges herein contained, the parties hereto agree as follows:

ARTICLE I—DEFINITIONS

For the purpose of this Agreement, the words and terms listed below shall have the following meaning:

A. The term "Licensed Products" shall mean _____.

B. The term "Patents" shall mean patents, utility models and design rights, by the Licensor and relating to the Licensed Products or the Technical Information, issued by any government and applications therefor that are filed prior to or during the term of this Agreement.

C. The term "Trademarks" shall mean the following trademarks and trademark applications in _____ owned by Licensor:

商　標　　　　　　　　　登記號碼

申請號碼

(4) 技術情報乃指有關於非授權產品之製造或使用等目的，所相關
之技術情報、訣竅、製圖、說明、販賣方法、製造之方式及秘
密、及原料之處理方法等現為授權人所有或控制者。

(5) 改良乃指技術情報之任何改良、發展或變更。

(6) 輸出領域乃指 （　）；但授權人於該等國家已指定獨家非授權
人或依授權人之判斷，對授權於該等國家亦有合理之切實性認
為市場混亂者，授權者有權要求撤出該所列之國家。

(7) 生效日乃指本契約依（法律）（政府）核准時，或依（法律）
之（政府）核准時，於前述政府最後核准時為準，且其核准之
形式與實際為當事人所接受。

第二條　授權
　1. 授權人之授權被授權人利用授權產品製造有關之專利或商標，
其權利之使用僅限於 （　）且其銷售亦僅限於 （　）輸出領域。
除 （　）及輸出領域之國家外，被授權人不得將授權之產品輸
出或銷售以供輸出。

Trademark   Registration Number

Application No.

D. The term "Technical Information" shall mean presently existing technical information, know-how, drawings, specification, shop methods, processes, formulas and secrets of manufacture, and treatment of materials presently owned or controlled by Licensor at the date hereof, pertaining to the manufacture and use of the Licensed Products.

E. The term "Improvement" shall means any improvement, development, or modification of the Technical Information.

F. The term "Export Territory" shall mean _____; provided, however, that Licensor has the right to withdraw any of the above enumerated countries in the event an exclusive franchisee has been appointed in such country or sales by Licensee in such country is in Licensor's judgment faced with a reasonable certainty of disturbing the market, such as by the possibility of dumping by Licensee's customers.

G. The term "Effective Date" shall mean the date on which the approval by the _____ Government of this Agreement required by _____ law or the approval by the _____ Government of this Agreement required by _____ law, whichever is later, shall have been obtained in such form and substance as are acceptable to the parties.

ARTICLE II—LICENSE

A. Licensor hereby grants to Licensee a non-exclusive right to use the Patents or the Trademarks in connection with the manufacture of the Licensed Products in _____ and sale thereof in _____ and the Export Territory. No Licensed Products shall be exported, or sold to others for export, to countries other than _____ and those in the Export Territory.

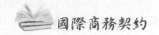

2. 授權人於契約生效日後，應將（　）提供予被技術情報授權人，以供被授權人為第二條第 1 項之目的而為使用。

第三條　技術協助

1. 於契約生效日後，授權人應迅速將（　）之完整資料技術情報交於被授權人。再者，授權人應提供被授權人於要求且以當事人合意之利用，將有關（　）技術情報之任何畫圖及文件交予被授權人。所有文件應以英文製作之，且其所使用之尺寸、重量及標準須為（　）常用之方式。

2. 於被授權人請求後，授權人應派遣技術人員於被授權人（　）之工廠。被授權人應負擔派遣技術人員之費用，包括但不限於該等人員之薪水、福利稅、及於（　）與（　）兩地間之頭等旅行費用及生活費用。

3. 授權人經指導及訓練被授權人派往（　）之技術人員，以取得有關被授權物品之製造敘述及其他資訊。所有該被授權人技術人員之費用應由被授權人負擔，授權人應取得訓練該等人員之費用。

第四條　商標之授權

1. 被授權人所製造且銷售之被授權產品如具有商標者，其外表、構成及機能須符合授權人之說明及標準。除非被授權人將擬製造之被授權產品之原形準備且交予授權人核准者，被授權人不得使用授權人之商標。如授權人認為授權產品不符合授權人之標準或說明者，被授權人應立即停止使用該授權商品之商標。

B. Licensor shall disclose to Licensee promptly following the Effective Date, the (Technical Information) in order to permit Licensee to use the same as specified in Article II-A hereof, and only for such purpose.

ARTICLE III－TECHNICAL SUPPORT

A. Immediately following the Effective Date, Licensor shall transmit one complete set of copies of its (Technical Information) to Licensee. In addition, Licensor shall provide Licensee with additional copies of any drawings and documents relating to such (Technical Information) to Licensee upon request and at a cost to be agreed upon. All documents will be in the English language and will employ such measures, weights and standards as are in customary use in _____ .

B. Upon request of Licensee, Licensor shall send technical personnel to Licensee's factory in _____ . Licensee shall bear all of Licensor's costs for sending such technical personnel including without limitation the salaries of such personnel and fringe benefits, and shall furnish and pay for first-class travel between _____ and _____ and living expenses.

C. Licensor shall instruct and train technical personnel of Licensee sent to Licensor in _____ in order to acquire full manufacturing techniques and other information as to the Licensed Products. All expenses of such technical personnel shall be borne by Licensee, and Licensor shall be paid by Licensee a cost to be agreed upon for training such personnel.

ARTICLE IV－TRADEMARK LICENSE

A. The Licensed Products manufactured and sold by Licensee when using any of the Trademarks shall conform in external appearance, component parts and function to Licensor's specifications and standards. Licensee shall not use any of the Trademarks unless a model of each such Licensed Products shall first have been prepared and presented to Licensor and approved by Licensor as prototype. Licensee shall immediately discontinue use of

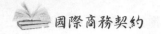

2. 對於授權人同意不申請登記或使用任何已授權人商標混淆之商標或商號名稱。

第五條　價金支付

1. 基於本契約授權及公開之對價，對授權人應支付其於契約有效期間銷售對授權產品之工廠銷售價格之百分之三與授權人。該銷售價格須與該授權產品工廠之銷售價格而減除營業稅、消費稅、運費、保險費及其他類似之費用。如被授權人就授權產品之使用、讓與或其他之處分為無償者，或以該產品使用、讓與或其他處分予獨立第三人價格為低者，該被授權產品之銷售價格應以該產品售予獨立第三人時之價格。

2. 依銷售價格所得之權利金應於本契約之期間於每季為計算，且於三月、六月、九月、十二月以後之六十日內予以交付。於每次交付價金時，對授權人應交予授權人有關於每季交付權利金之書面計算說明。

3. 被授權人保證依第五條一項所交付之權利金之額度，其每年不得少於下列之最低標準：

⑴ 第一年：零。

Trademarks on or in connection with any Licensed Products which Licensor determines do not conform with Licensor's standards and specifications.

B. Licensee agrees not to make application to register, or to use any Trademarks or trade means confusingly to any of the Trademarks.

ARTICLE V—PAYMENTS

A. As consideration for the License and disclosure provided for under this Agreement, Licensee shall pay to Licensor an amount equal to three percent (3%) of the ex factory selling price of the Licensed Products sold during the term of this Agreement by Licensee. Such selling price shall be computed ex factory on only the Licensed Products as are sold hereunder, and less all sales or excise tax, freight, insurance and other similar charges. In the event of use by Licensee or transfer or other disposition by Licensee of the Licensed Products without compensation or for a price lower than the price charged at the time of such use, transfer or other disposition to independent third party of such Licensed Products, then the selling price with respect to the Licensed Products so used, transferred or otherwise disposed of shall mean the price charged by Licensee at such time for sales of similar products to independent third party purchasers.

B. The royalty based on the selling price shall be computed quarterly during the term of this Agreement and shall be remitted within sixty (60) days after the end of each March, June, September and December. At the time of each such payment, Licensee shall render to Licensor a written statement showing the computation of the royalty hereunder payable for the calendar quarter for which such payment is made.

C. Licensee shall guarantee that the amount of royalty payment under Article V-A in each calendar year shall not be less than the following minimum royalty.

（ⅰ）For the first calendar year: None

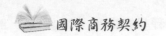

⑵ 第二年：（ ）美金（ ）元。

⑶ 第三年以後之各年度為（ ）元如未滿一年者，依其比率計
算之：美金（ ）元。

4. 如每年依第五條一項所支付之權利金低於第五條三項所規定之
最少額度時，該差額應於每年十二月後六十日內，或該契約終
止之年度內，支付予其差額及最後之權利金。依本契約應支付
予授權人之價金以美元計價之，該費率乃以匯兌時於（ ）之
官方銀行之售價計算之,該價金除非授權人另有以書面指示者，
其應支付於授權人之主營業所。

第六條　會計帳簿

被授權人應保存該授權產品銷售之正確會計帳簿及記錄。該帳簿
應於合理時間內，由授權人或其指示之代表人或簽證會計師予以
檢查。

第七條　租稅

依本契約支付予授權人之所得稅，應由授權人負擔之，且應依
（ ）之法律、協定、條約或其他（ ）及（ ）之合意為計算
之。如應支付予授權人之價金須予以扣繳，對授權人應交予該給
付及扣繳之證明書。

第八條　改良及授權

1. 授權人同意賦予被授權人使用由授權人所發展或製造之改良

(ii) For the second calendar year: $ _____

(iii) For each of the third and succeeding calendar year:

(Prorated on a per diem basis for a fraction of a year):

$ _____

D. In the event that the amount of royalty for any calendar year computed under Article V-A shall be less than the minimum royalty amount specified under Article V-C, the difference shall be paid together with the royalty payable within sixty (60) days after the end of each December, or, for the calendar year during which this Agreement shall be terminated. Under this Agreement amount paid shall be in United States Dollars, the amount of which shall be calculated at the official bank selling rate quoted in on the date of remittance, and to Licensor's principal place of business, unless otherwise directed in writing by Licensor.

ARTICLE VI—BOOK OF ACCOUNT

Licensee shall keep complete and accurate books of account, recording all orders for, production and sales of the Licensed Products. Such books shall, at all reasonable times, be open for inspection by Licensor or its designated representative or chartered accountants.

ARTICLE VII—TAXES

Any income tax imposed on payments to Licensor under this Agreement shall be borne by Licensor and shall be computed and paid in accordance with the laws of _____ and any convention, treaty or other agreement relating to taxation between _____ and _____ . In the event Licensee is required to withhold such tax from the amount paid to Licensor hereunder, and to pay the tax for the account of Licensor, Licensee shall provide Licensor certificates of such withholding and payment.

ARTICLE VIII—IMPROVEMENT AND GRANT BACK

A. Licensor shall grant to Licensee the right to use the Improvement made or

物，以便於授權人於（　）及（　）及出口領域為被授權人商品之製造。有關該等改良之專利或專利申請之權利，包括：使用由被授權人製造及發展出之改良物、非專屬及無償全世界之使用及其再授權等權利，應賦予授權人使用之。該等相互有關使用專利及專利申請之權利應於本契約期限內均得使用之，但當事人另有約定者不在此限。依本條項由授權人賦予被授權人之權利應依本契約第二條及第七條之條件辦理之。

2.為達成第七條一項之目的，當事人應使對造了解由其發展之改良或基於該改良所產出之專利或專利申請。

第九條　授權之禁止

除本契約另有訂定者外，授權人及被授權人如該授權乃法律所禁止或與第三人所訂立之契約所禁止者，則授權人對被授權人不負有下列之義務：

⑴公開任何發行、改良或新式樣，或

⑵授與專利或專利申請之授權。

第十條　侵害

1.授權人不保證有關專利、商標及（　）技術情報及改良將不侵害他人之權利。

2.如有關對被授權人之使用專利、商標、（　）技術情報或其他改

developed by Licensor for the manufacture of Licensed Products in ＿＿＿ and sale thereof in ＿＿＿ and the Export Territory. The right to use the Improvement made or developed by Licensee, and non-exclusive, world-wide royalty-free License, with the right of sublicense, of the patents and patent applications based on such Improvement shall be granted to Licensor without charge. Such reciprocal rights to use the patents and patent applications shall be for the life of this Agreement, unless otherwise agreed upon between the parties. The rights granted under this Article VIII-A by Licensor are also subject to all the terms and conditions of Article II and XII of this Agreement.

B. In order to attain the purposes mentioned in Article VII-A, each of the parties shall keep the other fully informed of the Improvement made or developed by it and of the status of patents and patent applications based on such Improvement.

ARTICLE IX－PREVENTION OF LICENSE

Anything in this Agreement to the contrary notwithstanding, neither Licensor nor Licensee shall be obligated to:

(a) disclose any inventions, improvements or new designs, or

(b) grant a License under any patent or application for patent, if a party obligated to so disclose or to grant such license hereunder is prohibited by law or by agreement with a third party from making such disclosure or from granting such license.

ARTICLE X－INFRINGEMENT

A. Licensor does not guarantee that the use of the Patents, Trademarks and the (Technical Information) and the Improvement will not infringe upon the rights of others.

B. In the event that any suits are instituted against Licensee for infringement of

良而致被請求侵害之訴訟發生時，被授權人應立即通知授權人此等訴訟，並准予授權人依其選擇以律師或顧問控制答辯之方法。如授權人承擔此等訴訟之進行，其答辯應由被授權人負擔費用，被授權人應完全為答辯而合作，且應提供其所握有之證據。被授權人於收到書面侵害之請求時，應立即通知授權人。

3. 如授權人與被授權人以書面同意有關專利、商標等授權之侵害救濟時，此等訴訟須以積極及謹慎之方法由被授權人提起並由授權人與被授權人共同進行之。就此等訴訟應由授權人與被授權人平均負擔其費用。因訴訟或和解所取得之費用應首先償還當事人之費用，如有剩餘時應平均分配予當事人。被授權人非經授權人之書面同意不得就訴訟為和解或其他之處分。縱被授權人就訴訟不為參加時，授權人仍得各別提起訴訟。

第十一條　不得異議
　　　　　對授權人同意不得以直接或間接之方法對專利或商標之有效提起異議或以其他方式為之。
第十二條　秘密之保持
　　　　　對授權人就依本契約所供應之技術情報應以秘密之方式保持，且僅得為本契約之目的而為使用。對授權人不得將技術情報洩

patents based solely on the practice by Licensee hereunder of the Patents, Trademarks, (Technical Information) or Improvements, Licensee shall promptly notify Licensor of the institution of such suit or suits and permit Licensor to control the defense thereof with counsel of its own selection. Licensor shall thereupon assume the control thereof, it being agreed that all expenses for such defense shall be for the account of Licensee. Licensee shall cooperate fully in the defense of such suits and to that end shall furnish all of the evidence in its control. Licensee shall also give Licensor immediate notice of receipt of any written claim of infringement.

C. In the event that Licensor and Licensee agree in writing that a suit should be brought for infringement of any of the Patents or Trademarks Licensed to Licensee hereunder, such suit shall be brought promptly thereafter and be vigorously and diligently prosecuted by Licensee subject to joint control of Licensor and Licensee. Licensor and Licensee shall each bear one-half of all of the expenses arising out of such suit. Any sum of money which may be collected in any suit by judgment or by settlement shall be used first to reimburse the parties for the expenses incurred by them with respect to such suit, and any balance then remaining shall be divided equally among the parties. Licensee shall make no settlement or other disposition of any such suit without the written consent of Licensor. Nothing herein shall preclude Licensor from bringing suit individually in the event Licensee elects not to participate in any such suit.

ARTICLE XI－NON-CONTESTABILITY

Licensee covenants not to contest, either directly or indirectly, the validity of any of the Patents or Trademarks, or aid others in doing so.

ARTICLE XII－MAINTENANCE OF SECRECY

Licensee shall handle in a confidential manner the Technical Information furnished by Licensor under this Agreement, and such information shall be

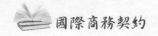

露於他人，但被授權人之受僱人對授權人產品之製造者不在此限，被授權人且須遵行授權人之意見而採取特別之防範以避免該技術情報不當或未經授權而洩露於他人。

第十三條　期間

1. 本契約之期間自生效日起為七年，然被授權人於終止日前六個月得以書面通知授權人將該期限延長為五年，惟有必要時，須經（政府）之認可。

2. 不論第十三條一項之規定，授權人得因被授權人自願或非自願之支付不能或破產、對債權人讓與其權利之情形，以書面通知授權人終止本契約，又如當事人違約時，且其違約於當事人以書面通知對方九十日內未為改正者，當事人之任一方均得以書面終止契約。

3. 如被授權人為他人合併，或其控制之權益為他人所控制，且該他人所進行之業務與授權人構成競爭者，得依授權人之意見認為其影響授權人之營業，授權人得以書面通知被授權人終止本契約。

4. 依本契約第四條二項、第十條與第十一條所規定之義務，得於契約終止後繼續有效。

used solely for the purpose stated herein. Licensee shall not disclose any of the (Technical Information) to others than those of employees who are actually engaged in the manufacture of the Licensed Products, and take such special precautions as in Licensor's opinion may be necessary to prevent any inadvertent unauthorized disclosure of any of the (Technical Information).

ARTICLE XIII—TERM

A. The term of this Agreement shall be from the Effective Date and shall remain effective for seven (7) years; provided, however, that Licensee may, by written notice to Licensor not less than six (6) months prior to such termination date, extend the term of this Agreement for an additional period of five (5) years, subject to validation by the _____ Government where necessary.

B. Notwithstanding the provisions of Article XIII-A, Licensor may terminate this Agreement forthwith upon written notice to Licensee in the event of the latter's insolvency or bankruptcy either voluntary or involuntary, assignment for the benefit of creditors, and either party may terminate this Agreement forthwith by written notice to the other in the event of a breach by the other of any of the provisions contained herein if such breach remains unrectified for a period of ninety (90) days following written notice of the breach.

C. In the event the Licensee should have been merged into, or the controlling interest thereof should have been controlled by, and the party is engaged in business competition with that of the Licensor or who by any other reason is feared, in the reasonable judgment of the Licensor, adversely to affect the Licensor's business, the Licensor shall have the option to terminate this Agreement by a written notice to be Licensee.

D. The obligations provided for under Article IV-B, Article X and Article XI hereof shall survive the termination of this Agreement for any reason whatsoever.

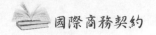

第十四條　不可抗力

　　1.如當事人之一方應不可抗力之事由，就非經給付之義務無法
　　履行時，當事人得以書面或電話技術通知他方該等事由，則
　　當事人於通知後就因不可抗力所生之債務應延緩履行，但該
　　等事由可以合理之方法除去者不在此限。

　　2.本契約所稱之不可抗力包括自然行為、罷工、停工、其他產
　　業上之阻礙、公敵行為、戰爭、封鎖、暴動、騷動、流行病、
　　山崩、打雷、地震、火災、暴風雨、水災、內亂、政府之規
　　範及其他非由當事人所得合理控制或盡其能力得以防止之事
　　由。然前述之罷工或停止之須為當事人依其裁量無法解決者，
　　且前述所謂當事人須盡其努力解決不可抗力之事由，並不包
　　括依當事人之判斷，就罷工或停止須滿足他造之需求。

第十五條　讓與

　　1.本契約非經對造當事人之同意不得讓與第三人，對授權人亦
　　不得將其授權再授權予第三人。

ARTICLE XIV－FORCE MAJEURE

A. In the event of either party being rendered unable, wholly or in part, by force majeure, to carry out its obligations under this Agreement other than to make payments of amounts due hereunder, it is agreed that on such party giving notice of such force majeure in writing or by cable to the other party with reasonable promptness after the occurrence of the cause relied on, then the obligations of the party giving such notice so far as they are affected by such force majeure shall be suspended during the continuance of any inability so caused but for no longer period, and such cause shall, so far as possible be remedied with all reasonable dispatch.

B. The term "force majeure" as employed herein shall mean acts of God, strikes, lockouts or other industrial disturbances, acts of the public enemy, wars, blockades, insurrections, riots, epidemic, landslides, lighting, earthquakes, fires, storms, floods, civil disturbances, governmental regulations, and any other cause whether of the kind herein enumerated or otherwise not within the reasonable control of the kind herein enumerated or otherwise not within the reasonable control of the party claiming suspension, all of which by the exercise of due diligence such party is unable to prevent; provided, however, that the settlement of strikes or lockouts shall be entirely within the discretion of the party having the difficulty, and that the above requirement that any force majeure shall be remedied with the exercise of due diligence shall not require the settlement of strikes or lockouts by according to the demands of the opposing party when such course is inadvisable in the discretion of the party having the difficulty.

ARTICLE XV－ASSIGNMENT

A. This Agreement shall not be subject to assignment by either party without the prior written consent of the other, neither is Licensee permitted to

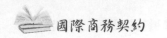

2.對授權人未經他造之書面同意而與他公司合併或與他公司合併而設立新公司時，授權人有權終止本契約。

第十六條　通知
本契約之通知須以書面為之，且對授權人應寄達（　）；對被授權人應寄達（　），或經契約當事人以書面通知他方之其他地址。

第十七條　仲裁
任何因本契約所生之爭執應由本契約之任何一方送交仲裁，且該仲裁應依據（　）之規則解決當事人之爭議，該仲裁得依前述之規則指定一或多數仲裁員。該仲裁應於（　）舉行之。該仲裁員之判斷應為終局之判斷。就該仲裁判斷之法院認可得由有管轄權之法院為之，或得由任一法院接受該判斷或得為執行之命令者。如有關該仲裁判斷之執行判決由管轄法院為之者，本契約當事人應放棄依準據法所得異議之權利。

第十八條　用語及準據法
本契約之作成與交付應以英文為之，當事人且同意於（　）成立本契約，且以（　）之法律為準據法。

sublicense the License granted hereunder to others.

B. In the event Licensee shall, without the prior written consent of the other, have merged into any other corporation or amalgamated with any other corporation to form a new corporation, Licensor shall have the right to terminate this Agreement.

ARTICLE XVI—NOTICE

Any notice required or permitted to be given hereunder shall be in writing, and in the case of Licensor, be addressed to ＿＿＿ ; and in the case of Licensee, be addressed to ＿＿＿ , or to such other addresses as any of the parties may from time to time designate by notice in writing to the other.

ARTICLE XVII—ARBITRATION

All disputes arising in connection with this Agreement shall be submitted to arbitration by any party and any dispute so submitted to arbitration shall be finally settled under the rules of the ＿＿＿ by one or more arbitrators appointed in accordance with such rules. Arbitration shall be held in ＿＿＿ . The award of the arbitrators shall be final. Judgment upon the arbitration award may be rendered in any court having jurisdiction thereof or application may be made to such court for a judicial acceptance of the award and an order of enforcement, as the case may be. In the event an action for judgment of execution is brought in a court of competent jurisdiction on the arbitration award or in the judgment rendered thereon, the parties waive all rights to object thereto insofar as permissible under the applicable laws.

ARTICLE XVIII—LANGUAGE AND LAW TO CONTROL

This Agreement has been executed and delivered in a text using the English language, which text, despite any translations into the ＿＿＿ language, shall be controlling. It is the intent of the parties hereto that this Agreement shall be deemed to have been made in ＿＿＿ , and shall be governed by the laws of ＿＿＿ .

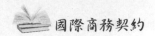

第十九條　全部條款

　　本契約乃當事人間之全部合意，本契約且應優於當事人前所交涉、表明、約束及其他合意之事項。

　　為證明本契約，當事人乃於前述之年月日作成本契約。

　　　　證明：　　　　　　　　　A 公司

　　　　　秘書　　　　　　　　　總經理

　　　　_____

　　　　證明：　　　　　　　　　B 公司

　　　　　秘書　　　　　　　　　總經理

　　　　_____

ARTICLE XIX—ENTIRE AGREEMENT

This Agreement constitutes the entire agreement between the parties, and supersedes all previous negotiations, representations, undertakings and agreements heretofore made between the parties with respect to the subject matter.

IN WITNESS WHEREOF, the parties hereto have caused this Agreement to be executed as of the day and year first above written.

Witness: A COMPANY, LTD.

  Secretary   By President

_____   _____

Witness: B COMPANY, LTD.

  Secretary   By President

_____   _____

§25 不動產抵押契約 (Real Estate Mortgage Agreement)

所謂之抵押契約乃指由抵押權人 (Mortgagee) 取得由債務人 (Debtor) 或第三人不移轉占有而供擔保之不動產,得就其所賣之價金受清償之權(參考民法第八六〇條)。不動產抵押契約一般均包括下列條款:

1.訂約之日期 (Date of agreement)。

2.當事人之姓名及地址 (Names and addresses of parties)。

3.抵押權人提供借貸事實之陳述 (A recital of the making of a loan by the mortgagee)。

4.抵押人提供抵押予抵押權人 (The mortgage of property by the mortgagor to the mortgagee)。

5.抵押之標的 (The property subject to mortgage)。

6.抵押之條件 (Terms of mortgage generally)。

7.對抵押權人之債務 (Amount owing to the mortgagee)。

8.債務之返還 (Repayment of the amount owing)。

9.抵押權人於債務人不履行時之權利 (Rights of mortgagee on default of debtor)。

10.抵押物孳息之處理 (Interest or other return from mortgaged property)。

11.抵押物之保管 (Care of mortgaged property)。

12.標的之保險 (Insurance of property)。

13.抵押物之處分 (Manner in which mortgaged property is to be disposed)。

14.抵押物之替換 (Right to substitute collateral)。

15.拍賣抵押物之權利 (Power to sell collateral at an auction)。

16.準據法 (Governing law)。

17.糾紛解決 (Dispute settlement)。

CONTRACT

1. Obj... **this Contract**
1.1. T...tomer shall order and the Executor shal...
consu...der the Technical Assignment (Ap...
Contra...ges) of the performance of...
1.2. Per...nment.
Technic...

2. Obligatio...**e Parties**
2.1. The Cu...r shall be obliged:
a) to pa...he work perform...
Contract...
b) to pro...mely all nec...
Executor;...o provide f...
c) if necess...be oblige...
2.2. The Executor...e of wor...under th...
a) to perform...he ful...
b) upon perfo...ntract a...
c) to perform v...
determined in th...

3. **Procedure of Work**
3.1. The Executor shall...m work...any third...
3.2. The Executor may el...conditions o...
however, subject to terms...pproval or no...
...ed for in this Contract...
...t required. ...ction of work and within the period...
...the Report on the performed work an...
...e the Customer for signing. ...days of the date of receipt of the Report an...
...tion, the performed work shall be d...
...t shall be argumented and incl...
...assignment. In such...

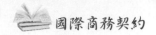

不動產抵押契約書

一、債務人（以下簡稱「債務人」）及擔保物提供人（以下簡稱「提供人」）
為擔保債務人對　　銀行（包括總公司及其各分支機構）（以下簡稱
「抵押權人」），在本抵押權設定契約書所定之本金最高限額以內，現
在及將來所負之債務（以下稱「該債務」），暨其利息、遲延利息、違
約金、行使及保全債權之費用以及因債務不履行所造成之損害賠償（以
下合稱「其他債務」），茲提供擔保物（詳列如附件一），設定第　順位
抵押權人。

二、有關該債務之清償期限、方法及提前到期等均應依抵押權人與債務人
所另簽訂之契約規定處理之。

三、利息按各筆債務契約所約定之利率計算，其給付方法照各個債務契約
之約定，逾期利息及遲延還本付息違約金亦照各筆債務契約所定之標
準計收。

四、債務人應接受抵押權人對該債務用途之監督及對債務人業務財務之稽
核，如需各項表報及資料，債務人應立即供給，但抵押權人並無監督
或稽核之義務。

MORTGAGE AGREEMENT

1. The debtor (hereinafter referred to as the "Debtor") and the party who provides the collateral(s) (hereinafter referred to as the "Party") in order to secure any and all indebtedness of the Debtor now and hereafter owed to _____ Bank (including the Head Office and all its branch offices, hereinafter referred to as the "Mortgagee") up to the maximum aggregate principal amount as specified herein (hereinafter referred to as the "Credit"), and interest, default interest, penalty and expenses incurred in connection with execution and enforcement of this Agreement, and the compensation for whatever damage which may occur as a result of failure in performing and observing any of the provisions hereunder (hereinafter referred to as the "Other Debts"), do hereby furnish the collateral (as listed in Attachment 1 hereof) and create a _____ preferred mortgage thereon in favor of the Mortgagee.

2. The date and manner of repayment of the Credit and acceleration of the Credit or portions thereof shall be governed by and determined in accordance within the provisions set forth in the relevant agreement executed by the Debtor and the Mortgagee.

3. Interest shall be calculated at the rate as provided for in the respective agreements. The payment thereof will be made in accordance with the provisions set forth in the respective agreements. The default interest and the penalty for delay in repayment of any principal installment and/or payment of any interest shall also be charged in accordance with the provisions set forth in the respective agreements.

4. The Debtor shall accept supervision as may be exercised by the Mortgagee with respect to the use of the Credit and auditing relating to the business and financial conditions of the Debtor; and shall immediately furnish various

五、債務人及提供人茲聲明所提供之擔保物完全為債務人及（或）提供人合法所有，無任何他人權利，亦未設定任何負擔，債務人及提供人不得就所提供之擔保物再設定抵押權。

六、債務人及提供人就擔保物應向主管機關辦理抵押權設定登記，並將所有登記證明文件及其他有關文件交抵押權人存執。擔保物之改良或添附應經抵押權人書面同意，因改良或添附或其他原因依法應為變更登記時，債務人及提供人應即辦理。

因前項規定所生一切規費及費用（包括合理之律師費用）均由債務人及提供人連帶負擔。

七、債務人及提供人應負責維護抵押物使其維持良好狀況並應以善良管理人之注意義務保管及使用擔保物，若抵押物因任何原因發生變動時，債務人及提供人應立即通知抵押權人。擔保物之稅捐、維護費及其他一切費用，均由債務人及提供人連帶負擔。若債務人或提供人未支付或怠於支付前項費用時，抵押權人除得依法律或本合約規定行使其權利外，並有權在法律許可範圍內，代付各該費用，如由抵押權人代付費用時，債務人及提供人應即連帶償還之。

kinds of reports and/or statements as may be required by the Mortgagee; however, the Mortgagee is under no obligation to exercise such supervision or auditing.

5. The Debtor and the Party do hereby represent and declare that the collateral provided hereunder is wholly and legally owned by the Debtor and/or the Party. No other party has any right over the collateral nor has any encumbrance, lien or security interest been created thereon and the Debtor and the Party will not create, nor cause to be created or suffer to exist any such encumbrance, lien or security interest.

6. The Debtor and the Party shall register the creation of mortgage on the collateral with the competent authorities in charge of registration thereof and forward to the Mortgagee all certificates and other documents relating to such registration. Any improvement or addition to the collateral shall be subject to the written consent of the Mortgagee. Registration should be forthwith amended in case of such improvement or addition to the collateral or for any other reasons required by laws and/or government ordinances.

All expenses and fees (including reasonable fees of legal counsel) incurred therefrom shall be jointly and severally borne by the Debtor and the Party.

7. The Debtor and the Party shall exercise due diligence as a good administrator in maintaining the collateral in good repair and using the collateral. If there shall occur any change in the present condition of the collateral, for whatsoever reason, the Debtor and the Party shall immediately give notice to the Mortgagee.

All taxes, levies, and imposition on, and all expenses incurred for the collateral including, but not limited to, maintenance shall be jointly and severally borne by the Debtor and the Party.

In case of the failure of the Debtor or the Party to pay any of the foregoing, the Mortgagee may, to the extent permitted under applicable laws, at its

八、擔保物應以抵押權人為優先受益人，向抵押權人認可之保險公司投保，保險金額及保險條款應商得抵押權人同意，所有保單及保費收據應交抵押權人收執，一切費用概歸債務人及提供人連帶負責。若債務人或提供人怠於投保或不為投保時，抵押權人除得依法律或本合約規定行使其權利外，並有權在法律許可範圍內，代付各該費用，如由抵押權人代付費用時，債務人及提供人應即連帶償還，如未即時償還，抵押權人得逕列入債務人於各個債務契約下之借款本金金額，於法律許可範圍內按本抵押契約書第三條規定之利率之上限計息，但抵押權人並無代為投保或代付保費之義務。如擔保物毀損滅失，保險公司以任何理由拒絕或遲延賠款，或賠款不足時，債務人及提供人應另行提供抵押權人許可之等值或較高價值之擔保物，或立即清償該債務暨其他債務。

九、擔保物若為土地及（或）房屋，其範圍包括現有及其後改良或添附之建築物、地上物、水利、花園、樹木及附屬之全部設備，包括自來水、

option and without affecting any rights or remedies hereunder or under applicable laws, advance the amount of the foregoing and shall be reimbursed by the Debtor or the Party, upon demand.

8. The collateral shall be insured with an insurance company acceptable to the Mortgagee with the Mortgagee as the preferred beneficiary. The insured amount and the terms and conditions of the insurance shall be subject to the Mortgagee's consent. All the insurance policies and premium receipts shall be surrendered to the Mortgagee for custody. All costs and expenses incurred thereunder shall be jointly and severally borne by the Debtor and the Party. In the event of failure of, or delay by, the Debtor or the Party to have the collateral insurred, the Mortgagee may at its option and without affecting any rights or remedies hereunder or under applicable laws, have the same insurred or have such premiums paid. The Debtor and the Party shall, upon demand, reimburse the Mortgagee for any insurance premium paid or advanced by the Mortgagee. If the reimbursement is not made immediately, the Mortgagee may add such amount to the outstanding unpaid amount owed by the Debtor under the respective agreements and may, to the extent permitted under applicable laws, charge interest at the maximum rate as provided for in Article 3 hereof. However, the Mortgagee is not obligated to have the collateral insured or to pay the premium on behalf of the Debtor or the Party. In the event that the collateral suffers any damage or loss and the insurance company refuses or delays for whatever reason to make indemnification or indemnification is insufficient, the Debtor and the Party shall provide other collateral, of equal or higher value, acceptable to the Mortgagee, or shall immediately repay the Credit and at the same time make full payment of the other Debts.

9. If the collateral is in the form of land and/or buildings, it shall include the buildings, superficies, water supply, garden, trees, and all the fixtures

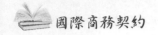

水井、瓦斯、電氣、冷暖氣、衛生設備等一切物件。

十、債務人或提供人或擔保物有下列情事之一者，抵押權人得要求債務人及（或）提供人另行提供抵押權之等值或較高價值之擔保物，或立即清償該債務暨其他債務，不受抵押權人所定清償期限之拘束，並得立即處分擔保物以抵償債務人之該債務及其他債務：

(1) 未依任一債務契約之約定按期還本利息，償付費用、稅捐或其他債務時；

(2) 債務人或提供人不履行本抵押權設定契約書或任一債務人與抵押權人或提供人與抵押權人間之任一債務契約之約定或債務人或提供人被發現在任一債務契約中作任何不實之陳述或聲明；

(3) 依公司法進行重整、合併（吸收或新設）或清算程序，或依破產法聲請和解或宣告破產，或受強制執行、假扣押、假處分或其他任何司法或行政上之處分；

(4) 受票據交換所拒絕往來處分或停止營業者；

(5) 出賣、出租、移轉或處分其全部或主要部分之營業或財產；

(6) 擔保物毀損、滅失、被留置、公用徵收、或其價值顯著減少者；或

appurtenant to the collateral, such as tap water, well, gas, power supply, air conditioning (cold and heating), sanitary equipment and other similar equipment, which are now or hereafter improved or added to the collateral.

10. Upon the occurrence of any of the following events on the part of the Debtor or the Party or with respect to the collateral, the Mortgagee may request that the Debtor and/or the Party provide other collateral, of equal or higher value, acceptable to the Mortgagee, or repay immediately the Credit and at the same time make full payment of the Other Debts without restriction on the repayment schedule determined by the Mortgagee. In addition, the Mortgagee may dispose of the collateral and apply the proceeds to the payment of the Credit and other Debts:

(1) default in making full payment of any indebtedness, expenses, taxes or any other sums when due;

(2) failure to perform or observe any provision of this Agreement or any other agreement between the Debtor and the Mortgagee, or between the Party and the Mortgagee or any statement or representation made therein shall prove to be incorrect;

(3) reorganization, consolidation, merger (whether it is the surviving company after merger), or liquidation under the Company Law, application for composition or petition for adjudication of bankruptcy under the Bankruptcy Law, or insolvent in any proceeding of compulsory execution, provisional seizure, provisional disposition or any other legal or administrative proceeding;

(4) discrediting by the Clearing House or cessation of business;

(5) sale, lease, transfer or disposition by other means of all or a substantial part of its business or properties;

(6) the collateral is damaged, lost, confiscated, appropriated for public use or its value is apparently diminished; or

⑺ 抵押權人主觀認為債務人就擔保物或融通資金運用不當或其他原因致有清償困難之虞者。

十一、抵押權人墊付依本抵押權設定契約書應由債務人及（或）提供人支付之任何費用，均應自墊付日起計入原本。

十二、債務人及提供人應將收取擔保物所生孳息、公用徵收補償金及其他任何收益之權利轉讓與抵押權人。

十三、本抵押權設定契約書以抵押權人所在地為履行地，如因本抵押權設定契約書涉訟時，以臺北地方法院為管轄法院。

十四、本抵押權設定契約書之效力應及於債務人及提供人之繼承人、受讓人、法定代理人、破產管理人、遺產管理人、公司重整人及公司清算人。

十五、本抵押權設定契約書規定未盡事宜，悉依中華民國有關法令、規章及銀行業慣例辦理。本抵押權設定契約書係中、英文並列，如有爭端，其在中華民國執行者應以中文為準，其在他國或地區內執行時，應以英文為準。

十六、於抵押權人與債務人另簽訂之相關合約下之所有融資、授信皆已完全終止且本契約所擔保之債務皆已完全清償後，抵押權人應即發給清償證明並協助債務人及／或提供人辦理抵押權塗銷登記，屆時，本契約並即終止。

(7) in the sole judgment of the Mortgagee, the collateral or the Credit is misappropriated by the Debtor or there shall exist any other reasons which may cause the payment hereof to become impaired.

11. Any expenses and fees which should be paid by the Debtor and/or the Party hereunder and which have been advanced by the Mortgagee shall be added to the Credit to be compounded from the date of advancement.

12. The Debtor and the Party hereby assign to the Mortgagee the right to collect the interest, compensation for appropriation of the collateral and any other income therefrom.

13. The location of the Mortgagee shall be the place of performance of this Agreement. In the event of any dispute arising herefrom, the Taipei District Court shall be the court having jurisdiction over the dispute.

14. This Agreement shall be binding upon the respective successor(s), assignee(s), legal representative(s), trustee(s), in bankruptcy receiver(s), administrator(s), reorganizer(s), and liquidator(s) of the Debtor and the Party.

15. For other matters not specifically covered hereby, the laws, government ordinances, regulations and established practices of banking institutions of the Republic of China shall govern. This Agreement is executed in both English and Chinese. If there is any dispute, the Chinese version shall govern if enforcement is sought in the Republic of China and the English version shall govern if enforcement is sought in any other territory or country.

16. Upon termination of the credit facilities extended under the relevant credit documentation executed by the Debtor and the Mortgagee and payment and repayment of the Other Debts in full, the Mortgagee shall immediately issue a certificate to the Party and/or the Debtor to such effect and assist the Party and/or the Debtor to deregister the mortgage created hereunder whereupon this Agreement shall terminate.

十七、債務人及提供人茲同意由其共同連帶負擔土地法第七十六條下之各
　　　項登記規費。

　　　　　　　　　　　　　（簽章）
　　　債務人

　　　地　　址

　　　　　　　　　　　　　（簽章）
　　　提供人

　　　地　　址

　　　銀行承諾及接受簽章

　　　日期：

17.The Debtor and the Party hereby agree that all registration fees under Article 76 of the Land Law of the R.O.C. shall be joint and several borne by the Debtor and the Party.

To:

_____

Debtor

_____

Address

_____

Party

_____

Address

Accepted and Agreed by:

_____

Date:

§26 一般智慧財產權授權契約 (General Intellectual Property License Agreement)

一般授權契約主要訂定下列條款：

1. 授權 (Grant of License)。

2. 權利轉讓 (Transfer of Rights)。

3. 終止 (Termination)。

4. 全部契約 (Entire Agreement)。

5. 修正 (Amendment)。

6. 個別性 (Severability)。

7. 放棄契約權利 (Waiver of Contractual Right)。

8. 準據法 (Applicable Law)。

CONTRACT

1. Obj
1.1. ...this Contract
...tomer shall order and the Executor shall
consu... ...der the Technical Assignment (Ap-
Contra... ...ges) of the performance of
1.2. Per... ...nment.
Technica...

...he Parties
2. Obligatio
2.1. The Cu... ...r shall be obliged
Contra... ...he work perfor...
a) to p... ...mely all nec...
b) to pro... ...o provide f...
Executor; ...be oblige...
c) if necess... ...under th...
2.2. The Executo... ...e of wo...
a) to perform... ...n the ful...
b) upon perfor... ...ntract a...
c) to perform... ...ntract a...
determined in th...

3. Procedure of Work
3.1. The Executor shall... ...m work t...
3.2. The Executor may e... ...any third ...
however, subject to terms... ...conditions o...
...ed for in this Contract... ...pproval or not...
...not required. ...the Report on the performed work and...
...tion of work and within the period in... ...to the Customer for signing.
...days of the date of receipt of the Report an...
...ion, the performed work shall be...
...shall be argumented and incl...
...assignment. In such

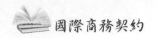

一般智慧財產權之授權契約

本授權契約（「本契約」）係於＿＿＿＿, 20＿＿＿＿, 由＿＿＿及＿＿＿簽署。

於本契約中, 授權之一方為「＿＿＿＿」, 被授權之一方為「＿＿＿＿」。

雙方同意如下:

1. ＿＿＿擁有＿＿＿（「＿＿＿」）。依本契約之約定, ＿＿＿授與＿＿＿專屬使用＿＿＿。

2. 本契約對任何一方受讓人具有約束力。

3. 本契約可由任何一方於 30 天前以書面通知另一方終止。

4. 本契約包含各方之前全部約定, 各方並沒有其他承諾或於任何其他口頭或書面契約之約定。本契約取代雙方之間之前任何書面或口頭契約。

5. 雙方得修正或修訂本契約, 該修正或修訂應以書面形式提出, 並經雙方簽署。

6. 如本契約之任何約定為無效或基於任何原因無法執行, 其餘條款應繼續有效且可執行。如法院認定本契約之任何條款無效或不可執行, 但如限制此約定之解釋, 此約定成為有效或可強制執行時, 則該約定條款應視為此約定之限縮適用。

GENERAL INTELLECTUAL PROPERTY LICENSE AGREEMENT

This License Agreement (this "Agreement") is made effective as of _____ , 20 _____ between _____ , of _____ , and _____ , of _____ .

In the Agreement, the party who is granting the right to use the licensed property will be referred to as " _____ ", and the party who is receiving the right to use the licensed property will be referred to as " _____ ".

The parties agree as follows:

1. _____ owns _____ (" _____ "). In accordance with this Agreement, _____ grants _____ an exclusive license to use the _____ .

2. This Agreement shall be binding on any successors of the parties.

3. This Agreement may be terminated by either party by providing 30 days' written notice to the other party.

4. This Agreement contains the entire agreement of the parties and there are no other promises or conditions in any other agreement whether oral or written. This Agreement supersedes any prior written or oral agreements between the parties.

5. This Agreement may be modified or amended, if the amendment is made in writing and is signed by both parties.

6. If any provision of this Agreement shall be held to be invalid or unenforceable for any reason, the remaining provisions shall continue to be valid and enforceable. If a court finds that any provision of this Agreement is invalid or unenforceable, but that by limiting such provision it would become valid or enforceable, then such provision shall be deemed to be written, construed, and enforced as so limited.

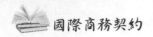

7. 任何一方未行使本契約之任何權利，不得解釋為放棄或限制任何一方之權利，本契約之約定應嚴格遵守。

8. 本契約以＿＿＿＿國之法律為準據法。

授權人：   被授權人：

＿＿＿＿＿＿＿＿＿＿＿＿   ＿＿＿＿＿＿＿＿＿＿＿＿

代表人：＿＿＿＿＿＿＿＿   代表人：＿＿＿＿＿＿＿＿

7. The failure of either party to enforce any provision of this Agreement shall not be construed as a waiver or limitation of that party's right to subsequently enforce and compel strict compliance with every provision of this Agreement.

8. This Agreement shall be governed by the laws of the State of ＿＿＿ .

Licensor: Licensee:

＿＿＿＿＿＿＿＿＿＿＿＿   ＿＿＿＿＿＿＿＿＿＿＿＿

By: ＿＿＿＿＿＿＿＿＿＿   By: ＿＿＿＿＿＿＿＿＿＿

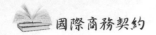

§27 專利讓與契約 (Assignment of Patent)

專利讓與契約主要訂定下列條款：

1. 發明人 (Inventor)。

2. 購買者 (Purchaser)。

3. 讓與之權利 (Rights to Assign)。

4. 發明人之擔保 (Warrant)。

CONTRACT

1. Obj... this Contract
1.1. ...tomer shall order and the Executor shal...
...der the Technical Assignment (Ap-
consu... ...ages) of the performance of ...
Contra...
1.2. Per... ...nment.
Technica...

2. Obligatio... ...e Parties
2.1. The Cu... ...r shall be obliged:
a) to pa... ...he work perfor...
Contrac...
b) to pro... ...mely all nec...
Executor; ...p provide 1...
c) if necess... ...be oblige...
2.2. The Executor... ...under th...
a) to perform... ...e of wor...
b) upon perfo... ...n the ful...
c) to perform... ...ntract a...
determined in th... ...ntract a...

3. Procedure of Work
3.1. The Executor shall ... ...m work ...
3.2. The Executor may e... ...any third...
however, subject to terms ... ...conditions o...
...led for in this Contract ... ...pproval or noti...
...t required. ...letion of work and within the period in...
...e the Report on the performed work an...
...the Customer for signing.
... ...n days of the date of receipt of the Report an...
...ion, the performed work shall be...
...k shall be argumented and incl...
...assignment. In suc...

專利讓與契約

〔＿＿＿＿〕為發明人，同意〔＿＿＿＿〕公司為買方，並以＿＿＿＿美元為代價，雙方茲同意：

發明人讓與買方、其繼承人及受讓人，美國專利編號 123456 相關之發明〔發明名稱〕之所有權利，包括申請專利或於全世界實施發明之相關權利。

發明人保證專利之所有權係完整、明確，發明人未知悉任何專利之主張及且發明人應以其費用保護其專利。

日期：＿＿＿＿＿＿＿＿＿

＿＿＿＿＿＿＿＿＿＿＿＿＿

簽名

＿＿＿＿＿＿＿＿＿＿＿＿＿

簽名

證人：＿＿＿＿＿＿＿＿＿＿＿

名稱／地址

ASSIGNMENT OF PATENT

[Name], referred to as INVENTOR, and [_____ Corporation], Inc., referred to as PURCHASER, in consideration of $_____ (& no/100 dollars) agree:

INVENTOR assigns to PURCHASER, and their heirs and assigns, all rights related to U.S. Patent Serial No. 123456, for an invention described as [name of invention], including rights to apply for a patent or to practice the invention worldwide.

INVENTOR warrants that the title to the patent is free and clear and that INVENTOR is not aware of any adverse claims thereto and shall defend the patent at its expense.

Dated: _____

_____
Signature

_____
Signature
Witness: _____
Name/Address

§28 商標使用授權契約 (Trademark License Agreement)

商標使用授權契約主要訂定下列條款:

1. 簡介 (Introduction)。

2. 契約之標的 (Subject Matter of Agreement)。

3. 授予之權利 (Grant of Rights)。

4. 授權商標之製造 (Manufacture of Trademarked)

5. 授權人商品檢驗 (Inspection of Merchandise by Licensor)。

6. 商標之使用 (Use of Trademark)。

7. 被授權人之商標保護 (Protection of Trademark by Licensee)。

8. 授權人之商標保護 (Protection of Trademark by Licensor)。

9. 被授權人之讓與 (Assignment by Licensee)。

10. 權利金 (Royalties)。

11. 權利金之支付 (Payment of Royalties)。

12. 權利金報表 (Royalty Statements)。

13. 被授權人保持之記錄 (Records to Be Maintained by Licensee)。

14. 授權人審計記錄之權利 (Licensor's Right to Audit Records)。

15. 被授權人不得損害商標之價值 (Licensee Not to Impair Value of Trademark)。

16. 契約期限 (Term of Agreement)。

17. 終止 (Termination)

18. 終止之效力 (Effect of Termination)。

19. 爭議之仲裁 (Arbitration of Disputes)。

20. 通知 (Notices)。

21. 準據法 (Applicable Law)。

CONTRACT

1. Obj... **this Contract**

1.1. ...tomer shall order and the Executor shall...
...der the Technical Assignment (Ap-
consu...
Contra...ges) of the performance of...
ment.

1.2. Per...
Technic...

2. Obligatio... **e Parties**

2.1. The Cu...r shall be obliged:
a) to p...the work perform...
Contract...mely all nece...
b) to pro...o provide f...
Executor,...be oblige...
c) if necess...under th...

2.2. The Executor...e of wor...
a) to perform...the ful...
b) upon perfor...ntract a...
c) to perform...ntract a...
determined in th...

3. Procedure of Work

3.1. The Executor shall...m work t...any third...
3.2. The Executor may e...conditions o...
however, subject to terms t...pproval or not...
...led for in this Contract...

...not required....tion of work and within the period i...

...on the performed work a...

...the Report on the performed work an...

...the Customer for signing.

...days of the date of receipt of the Report an...

...tion. the performed work shall be dec...

...t shall be argumented and incl...

...he assignment. In suc...

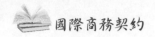

商標授權契約

1. 本契約於〔日期〕由〔授權人名稱〕(營業處所地址)與〔被授權人名稱〕(營業處所地址)簽署。

2. 授權人係「〔商標〕」商標所有權人,被授權人及授權人合意被授權人於(區域)使用商標。

3. 授權人授予被授權人於(區域)製造及銷售〔商品〕有關之商標使用權,被授權人承諾依本契約之條款於該區域中使用授權之商標。

4. 授權人得以自己之費用製造、散佈、出售商品。被授權人應依授權人不時所提供之規格、指示及進程,生產授權商標之商品。商品質量應獲授權人同意。

5. 授權人之授權代表,可於任何時間進入被授權人之處所,以確定被授權人生產之商品是否符合本契約第四條之約定。經授權人之要求,授權人須提交授權製造之商品樣品,以確定商品之質量是否獲授權人滿意。

6. 被授權人使用商標,為廣告、製造或銷售商品時,應註明授權人係商標所有權人。授權人將提供被授權人授權之包裝及標籤之樣本,除授權人另有書面指示,授權商品僅得使用該包裝及標籤。

TRADEMARK LICENSE AGREEMENT

1. Agreement dated [date], between [name], with principal offices at [address] (Licensor), and [name], with principal offices at [address] (Licensee).

2. Licensor is the proprietor of the trademark "[trademark]" (Trademark), and Licensee and Licensor wish to permit Licensee to use the Trademark in [description of territory] (Territory).

3. Licensor grants to Licensee the right to use the Trademark in the Territory in connection with the manufacture and sale of [description], the merchandise to which the Trademark applies (Merchandise). Licensee undertakes to use the Trademark in the Territory in accordance with the terms of this Agreement.

4. Merchandise by Licensee. Licensee may manufacture, distribute, and sell the Merchandise at Licensee's own cost and expense. Licensee shall use the Trademark only in connection with merchandise manufactured by it in accordance with the specifications, directions, and processes that Licensor furnishes to Licensee from time to time. The quality of the Merchandise shall be satisfactory to Licensor.

5. Licensor, by Licensor's authorized representatives, may at any time enter upon Licensee's premises to determine whether the Merchandise is being manufactured by Licensee in accordance with the provisions of Paragraph 4 of this Agreement. Upon Licensor's request, Licensee shall submit to Licensor's representatives samples of the Merchandise so that Licensor may determine whether the Merchandise's quality is satisfactory.

6. Licensee shall indicate Licensor's ownership of the Trademark whenever Licensee uses the Trademark in connection with the advertising, manufacture, or sale of the Merchandise. Licensor will furnish Licensee with samples of Licensor's packages and labels, and Licensee will use only those

7. 被授權人應遵守所有（區域）之法律，包括符合商標之標識要求。被授權人未能履行根據本款約定之義務，授權人可不通知被授權人且以被授權人之成本及費用，代表被授權人執行之。

8. 如授權人目前未於（區域）註冊商標，授權人將以自己成本及費用，於（區域）註冊商標。於本契約有效期內，授權人將保持於該區域商標有效註冊，將不容許被授權人以外之人於（區域）使用該商標。

9. 未經授權人之明確書面同意，本契約授予被授權人之權利，不得讓與權利予他人。被授權人得由他人生產商品，於此情況下，被授權人與製造商間之契約應經授權人之書面同意，授權人不得無理由擱置該契約。該契約應賦予授權人至被授權人之製造商處所檢驗之權，以確認商品是否按照本契約第四條之約定製造。

10. 被授權人應支付授權人出售商品總收入之百分之_____（____%) 為授權金。
〔另類條款〕
10. 被授權人將支付授權人出售之商品為每個單位百分之_____美元 ($____) 之授權金予授權人。
11. 權利金應按季計算，每個日曆季度後支付，不遲於該季之最後一天。第一期權利金應於首季終了期間〔日期〕支付。

packages and labels in connection with the Merchandise unless otherwise directed in writing by Licensor.

7. Licensee will comply with all the laws applicable to trademarks in the Territory including compliance with marking requirements. Should Licensee fail to perform Licensee's obligations under this Paragraph, Licensor may perform them on Licensee's behalf, without notice to Licensee and at Licensee's cost and expense.

8. If the Trademark is not presently registered in the Territory, Licensor, at Licensor's cost and expense, will register the Trademark in the Territory. Licensor will maintain the Trademark in the Territory during the term of this Agreement. Licensor, during the term of this Agreement, will not permit any one other than Licensee to use the Trademark in the Territory.

9. None of the rights granted to Licensee by this Agreement can be assigned to others without Licensor's express written consent. Licensee can have the Merchandise manufactured by others for Licensee. In such event, the agreements between Licensee and Licensee's subcontractors shall be subject to Licensor's written approval, which consent shall not be unreasonably withheld, and shall provide that Licensor may enter upon Licensee's subcontractors' premises to determine whether the Merchandise is being produced in accordance with the provisions of Paragraph 4 of this Agreement.

10. Licensee will pay Licensor a royalty of _____ percent (_____%) Licensee's gross receipts from the sale of the Merchandise.

[Alternative Paragraph]

10. Licensee will pay Licensor a royalty of _____ dollars ($_____) for each unit of the Merchandise sold by Licensee.

11. Royalties shall be computed quarterly and paid no later than the last day of the month following each calendar quarter. The first royalty payment shall be

12.權利金支付時，被授權人應提出被授權人因授權所生銷售之總收入及權利金計算之聲明。

13.被授權人須隨時保持本契約所涉及所有交易之詳實記錄。

14.授權人得以自己之費用，審計被授權人之書籍及記錄核實權利金報表。此權利應於記錄提出〔數字〕年後終止。費用審計得由註冊會計師或註冊會計師事務所，於正常營業時間，及〔數量〕天通知被授權人後為之。被授權人應使授權人亦取得類似之權利，審計再授權之簿冊及記錄。

15.被授權人承認授權人之專屬權、所有權、及商標之利益，並不得以任何方式損害或有致於損害任何授權人之權利、所有權及利益之任一部分。於使用該商標時，被授權人不得表示於授權人商標有任何所有權或為商標註冊。被授權人使用該商標，不得侵害授權人任何權利、所有權或利益。授權人於任何時間，於本契約期限內或終止後，未經授權人之事先書面同意，被授權人不得採用或使用任何文字或標記係相似或混淆之商標。

16.本契約之期限係〔數字〕年〔日期〕開始至〔日期〕終止。被授權人得於〔日期〕前予授權人書面通知延長本契約，即於〔日期〕開始及〔日期〕結束額外之〔幾〕年之期限。

for the period ended [date].

12. Each payment of royalties shall be accompanied by a statement showing the number of units of the Merchandise sold by Licensee during the period covered by the royalty payment, as well as Licensee's gross receipts for those sales and the computation of royalties due to Licensor.

13. Licensee shall at all times keep an accurate record of all transactions covered by this Agreement.

14. Licensor may, at Licensor's expense, audit Licensee's books and records for the purpose of verifying the royalty statements. This right shall terminate, with regard to each royalty statement, [number] years after the statement has been rendered. The audit must be made by a certified public accountant or certified public accounting firm, during regular business hours, and upon [number] days' notice to Licensee. In making any sublicensing agreements, Licensee shall obtain for Licensor a similar right to audit the sublicensee's books and records.

15. Licensee acknowledges Licensor's exclusive right, title, and interest in and to the Trademark and will not do anything that will in any way impair or tend to impair any part of Licensor's right, title, and interest. In connection with the use of the Trademark, Licensee will not represent that Licensee has any ownership in the Trademark or in its registration. Use of the Trademark by Licensee will not create any right, title, or interest in or to the Trademark in favor of Licensee. Licensee will not at any time, either during the term of this Agreement or after it has ended, adopt or use any word or mark that is similar to or confusing with the Trademark, without Licensor's prior written consent.

16. The term of this Agreement is [number] years starting on [date] and ending on [date]. Licensee may extend the term of this Agreement for an additional [number] years, starting on [date] and ending on [date], by giving Licensor

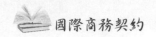

17.本契約於本條所載之任何事件發生時，即為終止或得為終止：

a.授權人不能清償、破產等。
如被授權人為債權人之利益讓與任何被授權人之資產或業務，或由受託人或管理人委任管理或進行被授權人之業務或事務，或被授權人於任何法律程序中裁定、自願或非自願破產，於此等情形，授權人得不另行通知或為法律行為，本契約所授予被授權人之權利即為終止。

b.被授權人未能遵守契約條款。
被授權人未能遵守本契約之任何約定，授權人得於不超過〔數量〕天書面通知被授權人終止本契約。但被授權人於通知期內改正，該通知應即為失效。

18.本契約因任何原因終止後，被授權人應停止使用該商標，並提供予授權人所有材料及文件，及出現商標之商品。除已進行之生產外，被授權人不得再為製造商品。被授權人於終止本契約〔數量〕個月內，可繼續出售已製成之任何庫存商品。被授權人須根據本契約之條款對該銷售支付權利金。如被授權人於本契約終止後〔數量〕個月內未售出庫存，該庫存須出售予授權人或以授權人之費用出售予授權人指定之某個人或實體。

19.除當事人另為合意，所有索賠、糾紛、本契約所產生之爭議及其他事項，應依美國仲裁協會之規則於〔城市〕仲裁。由仲裁人所為之裁決係終局

written notice no later than [date].

17. This Agreement will terminate or may be terminated upon the occurrence of any of the events set out in this Paragraph.

   a. Licensee's Insolvency, Bankruptcy, Etc.

    If Licensee makes any assignment of Licensee's assets or business for the benefit of Licensee's creditors, or if a trustee or receiver is appointed to administer or conduct Licensee's business or affairs, or it Licensee is adjudged in any legal proceeding to be either a voluntary or involuntary bankrupt, then the rights granted by this Agreement to Licensee shall cease and terminate without prior notice or legal action by Licensor.

   b. Licensee's Failure to Comply With Terms of Agreement.

    Should Licensee fail to comply with any provision of this Agreement, Licensor may terminate this Agreement upon not less than [number] days' written notice to Licensee. However, if Licensee corrects the default during the notice period, the notice shall be of no further force or effect.

18. Upon termination of this Agreement for any reason, Licensee will stop all use of the Trademark and will deliver to Licensor all material and papers, other than the Merchandise, upon which the Trademark appears. Licensee will also stop manufacture of the Merchandise except that work already in progress may be completed. Licensee may continue to sell any inventory that Licensee has on hand for a period of [number] months following the termination of this Agreement. Licensee must pay royalties on such sales to Licensor in accordance with the terms of this Agreement. If Licensor has unsold inventory at the end of the [number]-month period following termination of this Agreement, such inventory must be sold to Licensor or to a person or entity designated by Licensor at Licensee's cost.

19. All claims, disputes, and other matters in question arising out of this Agreement shall be decided by arbitration in [city] in accordance with the

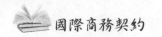

之決定，並依準據法，於任何有適當管轄權之法院為判決承認。

20.本契約所約定之所有通知須以掛號之郵件，並要求回執，以本契約所載或其他另一方給予之地址，送達通知另一方。

21.不論本契約於何地簽署，應根據〔國家〕之法律解釋本契約。

〔簽名〕
授權人
〔簽名〕
被授權人

rules of the American Arbitration Association then obtaining unless the parties mutually agree otherwise. The award rendered by the arbitrators shall be final, and judgment may be entered upon it in accordance with applicable law in any court having proper jurisdiction.

20. All notices required by this Agreement shall be given by registered or certified mail, return receipt requested, addressed to the party to whom notice is given at the address set out in this Agreement or such other address that the party to whom notice is given may have given the other party notice of.

21. This Agreement shall be interpreted in accordance with the laws of [state] regardless of where it is executed.

[signature]
Licensor
[signature]
Licensee

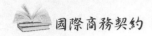

§29 存證信函 (Notice to Cease and Desist)

存證信函主要訂定下列條款：

1. 權利人及作品名稱 (Name)。

2. 停止使用 (Cease Use)。

3. 損害賠償 (Damages)。

4. 聯絡 (Contact)。

CONTRACT

1. Obj... ...his Contract

...tomer shall order and the Executor shal'...
...der the Technical Assignment (A...

1.1. ...
consu... ...ges) of the performance of...
Contra...
1.2. Per...
Technica... ...nment.

...e Parties

2. Obligatio...
2.1. The C... ...shall be obliged...
Contract... ...he work perform...
a) to pa...
b) to pro... ...mely all nece...
Executor... ...o provide f...
c) if necess... ...be oblige...
2.2. The Executo... ...under th...
a) to perform... ...e of wor...
b) upon perfo... ...n the ful...
c) to perform... ...ntract a...
determined in th...

3. Procedure of Work
3.1. The Executor shall... ...m work t...
...any third...
3.2. The Executor may e... ...conditions o...
however, subject to terms... ...pproval or noti...
...ed for in this Contract...
...t required. ...tion of work and within the period in...
...e the Report on the performed work an...
...the Customer for signing.
...days of the date of receipt of the Report an...
...tion, the performed work shall be de...
...shall be argumented and incl...
...by assignment. In su...

存證信函

日期：＿＿＿＿
對象：＿＿＿＿
名稱：＿＿＿＿
地址：＿＿＿＿
　　＿＿先生／女士：

作者（名稱）是著作權作品（名稱）之所有權人。我們已被告知台端使用著作權作品（名稱）。

根據「著作權法」，台端使用係非法。

依此，茲通知台端停止使用著作權作品，台端並應於 72 小時內交付非法使用之著作權作品。

著作權法訂有之大額侵權法定賠償及其他救濟措施。

如您需要任何額外之資訊，請以書面形式與我們聯繫。

敬告

＿＿＿＿＿＿＿＿＿＿＿＿

關於書面通知之通訊聯繫方式：

（聯繫地址）

NOTICE TO CEASE AND DESIST

Date: _____

To: _____

Name: _____

Address: _____

Dear _____ :

[Name] is the author of a copyright work entitled [Name]. We have been advised of your use of [Name].

Pursuant to the copyright act, such use is unlawful.

Accordingly, you are notified to cease use of the work, and to make arrangements to surrender all copies of the work within 72 hours.

The copyright code provides for substantial statutory damages and other remedies for infringement.

Should you require any additional information, please contact us in writing.

Sincerely,

_____

Contact for written communications regarding this notice:

[Address]

§30 營業秘密之保密契約 (Trade Secret Non-Disclosure Agreement)

按不論智慧財產權之授權契約、僱傭契約或其他商業交易，大多均訂有營業秘密之保密條款或保密契約。由於我國申請加入世界貿易組織 (WTO)，依智慧財產權貿易協定 (TRIPs) 之規範，我國乃訂有營業秘密保護法，此法律對營業秘密之保障有更詳細之規定，該種保密契約通常均訂有下列事項：

1. 訂約日期 (Date of Agreement)。
2. 當事人名稱及地址 (Names and Addresses of Parties)。
3. 訂約之緣由 (Recital of the Purposes of Agreement)。
4. 保密之事項 (Matters Subject to Non-disclosure)。
5. 被告知人之義務 (Duties of Confidant)。
6. 當事人之簽名 (Signature of Parties)。

CONTRACT

1. Obj... this Contract
1.1. ...tomer shall order and the Executor shal'
consu... der the Technical Assignment (Ap...
Contra... ges) of the performance of
1.2. Per... nment.
Technic...

2. Obligatio... ne Parties
2.1. The Cu... shall be obliged:
a) to pa... he work perform...
Contract... mely all nece...
b) to pro... provide f...
Executor... be oblige...
c) if necess... e of wor...
2.2. The Executor... under th...
a) to perform... the ful...
b) upon perfor... ntract a...
c) to perform... ntract a...
determined in th...

3. Procedure of Work
3.1. The Executor shall... m work t...
3.2. The Executor may e... any third...
however, subject to terms... conditions o...
...led for in this Contract... pproval or noti...
...t required.

...ction of work and within the period i...
...e Report on the performed work an...
...the Customer for signing.
...days of the date of receipt of the Report a...
...ion, the performed work shall be d...
...ion, the performed work shall be argumented and incl...
...assignment. In suc...

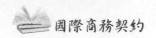

營業秘密之保密契約

　　按×××（告知人姓名）擬告知×××（被告知人姓名）有關×××（事實細節，以下簡稱營業秘密），以為雙方之共同商業目的。

　　被告知人茲同意遵守下列事項：

　1.保護營業秘密之機密性，且不洩漏予第三人。

　2.僅為告知人及被告知人之共同商業利益，且不得僅為被告知人之商業目的。

　　被告知人且接受如本約定書之部分內容為無效，將不影響本約定書之其餘部分。

　　於××年××月××日約定並蓋印。

_____　　　　_____
　　告知人簽名　　　　　　　　　　　被告知人簽名

TRADE SECRET NON-DISCLOSURE AGREEMENT

Whereas _____ (the "Confider") is
(confider name)

prepared to disclose certain trade secrets relating to _____

_____

(insert details)

(the "trade secrets") to _____ (the "Confidant")
(confidant name)

for mutual business purposes;

The Confidant hereby covenants:

1. to protect the confidentiality of the trade secrets and not to disclose them to any third-party;

2. to exploit the trade secrets only for the Confider's and the Confidant's mutual business purposes and not to exploit them for the Confidant's sole business purposes.

The Confidant accepts that, if any part of these covenants is void or unenforceable for any reason, it shall be severed without affecting the validity of the balance of the covenants.

Given under seal on _____
(date)

_____   _____
(Confider signature)   (Confidant signature)

§31 不可撤回之信託 (Irrevocable Trust)

按信託可分為撤回及不可撤回，其主要的差異在於委託人之權限，及未來課稅之標準，通常將財產讓與子女且為避免課稅之問題者，均設定不可撤回之信託，其主要之條款如下：

1. 信託財產之管理 (Administration of Trust Estate)。

2. 子女之信託 (Child's Trust)。

3. 處分 (Disposition)。

4. 受託人 (Trustee)。

5. 受託人之財務權限 (Financial Power)。

6. 行政管理權利 (Administrative Powers)。

7. 人壽保險 (Life Insurance)。

CONTRACT

1. Obj... this Contract
1.1. ...tomer shall order and the Executor shal'
consu... der the Technical Assignment (Ap...
Contra... ...ges) of the performance of
1.2. Per... ...nment.
Technica...

2. Obligatio... ...e Parties
2.1. The Cu... ...T shall be obliged:
a) to pa... ...he work perform
Contrac... ...mely all nec...
b) to pro... ...o provide f...
Executor; ...be oblige...
c) if necess... ...under th...
2.2. The Executor... ...e of wor...
a) to perform... ...h the ful...
b) upon perfor... ...ntract a...
c) to perform w... ...ntract and
determined in th...

3. Procedure of Work
3.1. The Executor shall... ...rm work t...
3.2. The Executor may en... ...any third p...
however, subject to terms a... ...conditions o...
...ded for in this Contract... pproval or noti...
...not required. ...tion of work and within the period 1...
...the Report on the performed work an...
...the Report on the performed work a...
...the Customer for signing.
...days of the date of receipt of the Report...
...tion, the performed work shall be...
...shall be argumented and incl...
...assignment. In suc...

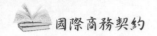

不可撤回之信託

當事人 A 住所＿＿＿茲以附件之財產轉讓於住所位於 ＿＿＿ 之 B 公司，委任 B 公司為其財產 ("C") 之受託人，該財產、組織及其他累積之資產將以下列方式管理：

第一條　不可撤回之信託
　　　　A 茲放棄有關本文件或任何信託得以變更或撤回之權利。

第二條　信託財產之管理
　　　　於本契約簽署之日及下述之日期止，受託人必須將有關信託之收入或本金分配予 A 之繼承人，繼承人亦得以書面向受託人請求該等分配。

第三條　子女信託
　　　　子女信託以以下之方式管理：
　　　　於本契約開始至終止日，基於健康、補助及教育之需求，受託人得決定分配予委託人之子女或其他繼承人有關本信託之收益或本金。

第四條　最終分配
　　　　於本契約因期限屆滿而終止時，受託人得將未依前述條款有效處分之信託本金，由受託人決定對慈善機構為分配之。

IRREVOCABLE TRUST

I, A, of _____ (place) _____ , hereby transfer to B of (place) _____ , as trustee of the "C", the property identified on the attached Schedule of Property, which property and all additions, investments, and accretions shall be administered upon the following terms:

ARTICLE I　Irrevocable Trust

I waive irrevocably all rights, power and authority to amend or revoke this instrument or any trust hereby evidenced.

ARTICLE II　Administration of Initial Trust Estate

Commencing on the date of this instrument and until the division date, the trustee shall distribute to each of my descendants who are living from time to time such amounts of the net income and principal of the trust as the descendant may demand in writing delivered to the trustee.

ARTICLE III　Child's Trust

Each trust named for a child of mine shall be administered as follows:

Commencing as of the division date and until the termination date (defined later in this Article), the trustee shall distribute to any one or more of the child and his or her descendants living at the time of the distribution as much of the net income and principal of the trust, even to the extent of exhausting principal, as the trustee determines from time to time to be required for the health, support and education of the child and his or her descendants;

ARTICLE IV　Ultimate Disposition

Upon termination of a trust at the end of its stated term under this instrument, the trustee shall distribute any trust principal not otherwise effectively disposed of by the foregoing provisions of this instrument to such one or more organizations selected by the trustee, each of which is, at the time of payment, a charitable organization, and in such proportions among such

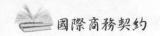

第五條　受託人條款

　　　　受託人之指定人得另指定其他受託人或受託人的繼受人，該等指
　　　　定必須以書面為之，其得以特定事件發生或到期日為生效期，其
　　　　受託人受託有效期間得為任何期間或為任何事件，得為定期或不
　　　　定期、限定或一般目的，得指定一人或數人，使指示事項應於文
　　　　件中載明之，受託人指定人得於被指定人接受之前撤回之，除指
　　　　派文件有明定者外，該指定得由其他受託人指定人撤回之，如指
　　　　定或撤回之表示有相衝突時，以較後之指示為準，受託人之指定
　　　　人應以信賴義務為信託之受益人最佳利益執行之。

第六條　財務權利

　　　　除法律賦予受託人之權利規定外，受託人得依其裁量執行本契約
　　　　信託之相關權利，受託人得將移轉之財產保存時日，該財產包括
　　　　合夥之利益、有限公司之權益等，縱受託人無法正當購買該信託
　　　　財產作為信託投資，及任何財產之保存違反投資分散原則，仍得
　　　　不負財產價值損失或貶值之責任。

第七條　管理之權利及規則

　　　　信託之管理應以下列原則為之：如得受領信託利益或本金之受益
　　　　人因故無能力時，受託人得以下列方式處理之：1.將該分配直接
　　　　分配予受益人或其監護人，或依據統一未成年法將財產交於符合

organizations as the trustee shall decide.

ARTICLE V　Trustee Provisions

The Trustee Appointer at any time may appoint any one or more Qualified Appointees as additional or successor trustees. Any appointment of an additional or successor trustee hereunder shall be in writing, may be made to become effective at any time or upon any event, may be for a specified period or indefinitely, may be for limited or general purposes and responsibilities, and may be single, joint or successive, all as specified in the instrument of appointment. The Trustee Appointer may revoke any such appointment before it is accepted by the appointee. An appointment may be revoked by a subsequent Trustee Appointer unless the instrument of appointment specifies otherwise. In the event that two or more instruments of appointment or revocation by the same Trustee Appointer exist and are inconsistent, the latest by date shall control. The Trustee Appointer shall act only in a fiduciary capacity in the best interests of all trust beneficiaries.

ARTICLE VI　Financial Powers

In addition to all powers granted by law, the trustee shall have the following powers with respect to each trust held under this instrument, exercisable in the discretion of the trustee, to retain for any period, without liability for loss or depreciation in value, any property transferred to the trustee, including partnership interests of any kind or limited liability company membership interests, even though the trustee could not properly purchase the property as a trust investment and though its retention might violate principles of investment diversification.

ARTICLE VII　Administrative Powers and Rules

If a beneficiary eligible to receive income or principal distributions is disabled at the time of distribution, then the trustee may, without further responsibility, either (i) make those distributions to the beneficiary directly, to

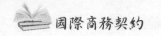

資格之個人、信託公司為受益人之保管人，2.為受益人之利益將財產為使用，受託人依本條款之決定應以誠信原則為之，並對所有關係人有最終之效力。

第八條　人壽保險

人壽保險為信託之收入與本金正當投資，除本契約另賦予受託人權利外，受託人得有完全之權限就保險契約收取、取得、或保存其權益，該等保險包括一人、數人、直接或間接為受託人適當之分配，受託人並得以信託之本金或收入為下列行為支付保險費，以其他個人或適當之公司為分配、將定期之人壽保險轉為整體制人壽保險、及行使任何權利包括但不限於變更受益人，將該保險單借貸金額、設定擔保、將保險單做適當之處分、降低保險費以購買其他之保險、使得保險金用以支付保險費、轉讓保險單為已支付型保險單等。

a lawful guardian of the beneficiary, or to a qualified individual or trust company designated by the trustee as custodian for that beneficiary under an applicable Uniform Transfers to Minors Act or similar law, or (ii) expend that distributable property for the benefit of the beneficiary in such manner as the trustee considers advisable. Determinations made by the trustee under this paragraph in good faith shall be conclusive on all persons.

ARTICLE VIII　Life Insurance Provisions

　　Life insurance policies are proper investments of trust income and principal. In addition to the other powers granted to the trustee, the trustee at all times and in all events, and acting alone in the sole discretion of the trustee, is specifically authorized, but not obligated, to receive or acquire, and to retain, any one or more policies of insurance, including policies on my life, the life of any other person, or the joint lives of any two or more persons, directly or indirectly in any arrangement deemed appropriate by the trustee; to pay such premiums as are required or permitted to be paid thereby out of trust income or principal; to enter into split-dollar or premium sharing arrangements with any individual (including me and my spouse) and any appropriate corporation or other entity; to split any policy into two or more policies; to convert any policy of term life insurance to whole life insurance; and to exercise any and all rights, powers, discretions, elections, options, privileges, and other incidents of ownership with respect to any such policy available to the trustee as owner, including without limitation the rights to change the beneficiary, to borrow money for any purpose (from the issuing company, the banking department of any corporate trustee acting hereunder, or others) using the policy as security, to select optional modes of settlement, to allow dividends to remain with the insurance company or to use them to reduce premiums or to purchase

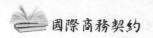國際商務契約

　　　　本契約於 ＿＿＿ 年 ＿＿＿ 月 ＿＿＿ 日於 ＿＿＿ 簽署之。

A   B

_____   _____

additional insurance, to allow cash values to be used to pay premiums, to surrender any such policy for its cash value, or to convert any such policy to a paid-up policy.

      Entered into in ＿＿＿ on ＿＿＿, 20 ＿＿＿.

A   B

＿＿＿＿＿＿＿＿＿＿＿＿＿＿＿＿＿   ＿＿＿＿＿＿＿＿＿＿＿＿＿＿＿＿＿

§32 宣言信託 (Declaration of Trust)

　　按所謂之信託乃指「委託人將財產權移轉或為其他處分，使受託人依信託人本旨，為受益人利益或為特定目的，管理或處分信託財產之關係」（信託法第一條）。信託原則上以契約為之，但如為公益信託，則得以單獨行為之方式，即「法人為增進公共利益，得經決議對外宣言自為委託人及受託人」（同法第七一條第一項）。該種宣言信託須由委託（即受託人）為宣言，其應包括下列事項：

　　1.宣言之日期 (Date of Declaration)。

　　2.委託人之名稱及地址 (Name and Address of Settlor)。

　　3.信託財產 (Trust Property)。

　　4.信託目的 (Purpose of Trust)。

　　5.受益人 (Beneficiary)。

　　6.委託人處分之限制 (Restriction of Disposition by Settlor)。

CONTRACT

1. Obj...
...tomer shall order and the Executor shall'
...der the Technical Assignment (A...
1.1. ...
consu...
...ges) of the performance of
Contra...
1.2. Per...
Technic...ment.

...his Contract
...e Parties
...shall be obliged:

2. Obligatio...
2.1. The Cu... he work perform
  a) to pa...mely all nec...
  Contract...
  b) to pro... provide f...
  Executor; ...be oblige...
  c) if neces... under t...
2.2. The Executor ...e of wor...
  a) to perform ...n the ful...
  b) upon perfo... ntract a...
  c) to perform ...ntract a...
  determined in th...ntract a...

3. Procedure of Work
3.1. The Executor shall ...rm work t... any third...
3.2. The Executor may e...
however, subject to terms ...conditions o...
...ed for in this Contract ...pproval or not...
...required ...ation of work and within the period i...
...the Report on the performed work and...
...the Customer for signing.
...days of the date of receipt of the Report a...
...on, the performed work shall be argumented and incl...
...assignment. In such...

宣言信託

　　本宣言信託乃由×××（受託人姓名，以下簡稱受託人）為×××（受益人姓名，以下簡稱受益人）而設定。受託人茲鄭重聲明就×××（財產明細，以下簡稱財產），乃僅為受益人利益而予信託方式持有。

　　受託人且對受益人承諾：
　　1.除交付予受益人外，非經受益人指示及同意，不得對財產為任何形式之處理；且
　　2.就所持有之財產，乃為受益人利益而受領金錢。

　　於××年××月××日簽署並用印。

　　　　　　　　　　　　　　　　　　　　　　＿＿＿＿＿＿＿＿＿＿＿＿＿
　　　　　　　　　　　　　　　　　　　　　　　　　受託人簽名

DECLARATION OF TRUST

This declaration of trust is made by (1) _____ (the "Trustee") in
(trustee name)

favor of (2) _____ (the "Beneficiary").
(beneficiary name)

The Trustee solemnly declares that he holds _____ (the
(list details)

"Property") in trust solely for the benefit of the Beneficiary.

The Trustee further promises the Beneficiary:

1. not to deal with the Property in any way, except to transfer it to the Beneficiary, without the instructions and consent of the Beneficiary; and,

2. to account to the Beneficiary for any money received by the Trustee in connection with holding the Property.

Given under seal on _____
(date)

_____
(Trustee signature)

主要參考書籍

一、中文部分

馮大同，國際貨物買賣法（北京：對外貿易教育出版社，一九九三年）。

蔣晃康等主編，國際民事商事公約與慣例（北京：中國政法大學出版社，一九九三年）。

董占東主編，最新經濟合同法全書（北京：國際文化出版公司，一九九三年）。

張錦源，英文貿易契約實務（臺北：三民書局，一九七七年）。

李永然主編，契約書製作範例（臺北：五南書局，一九九二年初版二刷）。

金融人員研訓中心，銀行定型化契約之研究（臺北：金融人員研訓中心，一九九五年）。

陳冲，信用狀基本法律理論（臺北：金融人員研訓中心，一九九三年五版）。

楊培塔，當前銀行押匯問題之研究（臺北：金融人員研訓中心，一九九五年三版）。

陳春山，契約法講義（臺北：瑞興圖書公司，一九九五年）。

二、日文部分

我妻榮，債權各論（東京：岩波書店，昭和四八年）。

伊藤進等，契約法（東京：崇陽書房，一九九〇年）。

北川善太郎，現代契約法（東京：商事法務研究會，昭和五一年）。

山本進一等編，債權各論（東京：青林書院新社，一九七八年）。

遠藤原之助，債權各論講義（東京：文久書局，一九七九年）。

田山輝明，契約法（東京：成文堂，昭和六〇年）。

松阪佐一，契約法大系（東京：有斐閣，昭和三八年）。

大野文雄、矢野正則，契約全書（東京：青林書院新社，昭和四八年）。

並木俊守，英文契約書作成の手引（東京：中央經濟社，昭和四七年）。

岩崎一生，英文契約書（東京：同文館，一九八一年）。

三、英文部分

The American Law Institute, *Uniform Commercial Code* (St. Paul, Minn.: West Pub., 1994).

Essler, Friedrich, et al., *Contracts* (Boston: Little, Brown & Company, 1970).

Farnsworth, E. Allan, et al., *Cases and Materials on Contracts* (Mineola, N.Y.: The Foundation Press, Inc., 1972).

Greenhill, Amelia C., *Forms of Agreement* (New York: Mattaew Bender, 1984).

Jacobs, Arnold, *Manual of Corporate Forms for Securities Practice* (New York: Clark Boardman Callaghan, 1992).

March, P. D. V., *Comparative Contract Law* (Hampshire: Gower Publishing, 1994).

Mueller, Addison, *Contract Law and its Application* (New York: The Foundation Press, Inc., 1977).

Penn, G. A., *Law & Practice of International Banking* (London: Sweet & Maxwell, 1987).

Sanderson, Steve, ed., *Standard Legal Forms and Agreements* (North Vancouver: Self-Counsel Press, 1994).

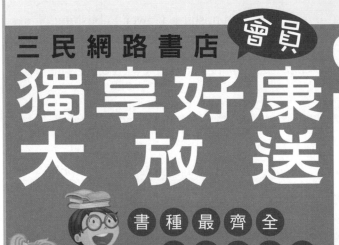